USEFUL PLANTS OF THE SOUTHERN CALIFORNIA MOUNTAINS

2nd Edition

USEFUL PLANTS OF THE SOUTHERN CALIFORNIA MOUNTAINS

2nd Edition

By
Ray S. Vizgirdas
Edna M. Rey-Vizgirdas

First Printing: 2018

ISBN # 978-0-359-22054-0

Mountainforaging@gmail.com

Disclaimer

While this book documents the many uses of wild plants found within the mountains of Southern California, the publisher and author disclaim any liability for injury that may result from following the instructions for collecting, preparing or consuming plants described in this guide. Efforts have been taken to assure the descriptions and drawings of plants represented are accurate representations of the family, genus, and species noted. It should be understood that growth conditions, improper identification, and varietal differences, as well as an individual's own sensitivity or allergic response can contribute to a hazard in sampling or using a plant. Furthermore, the reader is encouraged to seek the assistance from experienced botanists in identifying any of the plants discussed in this book.

To my old friend

Richy Reivydas

PREFACES AND ACKNOWLEDGMENTS

Preface to this Second Edition

In 2003, the first edition of *Useful Plants of the Southern California Mountains* was published by the San Bernardino County Museum Association, and only couple of thousand copies were printed, which eventually sold out. In the last ten or so years, I received hundreds of requests to get a copy or somehow make the book available again. During that time, I've been seriously thinking to totally update and revise the manuscript because of all the changes in plant taxonomy since that time and due to the release of the 2012 Jepson manual. Obviously, to accomplish such a task would take some time and money; but the email requests just kept on coming in.

Therefore, as a way to start the process, I decided to publish a second edition of the book – making minor edits, adding some new illustrations and photographs, and making it "pocket-sized' (as requested by so many people). So, here is the second edition of the *Useful Plants of the Southern California Mountains*. I believe it is still very useful to the outdoor adventurer wanting a general introduction to the ecology and ethnobotany of the plants found in the Southern California mountains. My thanks again to Ms. Jennifer Reynolds at the museum for allowing this new edition to move forward. This edition should serve as a placeholder as I make plans for the "completely revised and updated" edition in the next couple of years.

For this edition, I am very thankful to the following individuals for their support: Dr. Aras Mattis,

Dr. Daiva Mattis, Ms. Angela Staup, Ms. Aida Merrill, and Ms. Lina Ruplenas and her family.

Preface from the First Edition

Why would someone from Idaho be writing a book on the useful plants of Southern California Mountains? Very simply, a significant portion of my life has been associated with these mountains. Even as I explore the Northern Rockies today, many of my fondest memories with plants and outdoor living relate back to my many years of exploring the San Bernardino Mountains - specifically, Holcomb Valley. It is here that I explored and learned how to appreciate life.

This book is meant to be a field guide to the useful plants found in the mountains of southern California. It is intended to be used by wildlife enthusiasts, naturalists, city dwellers, hunters, backpackers, fishermen, prospectors, campers, vegetarians, and students of wilderness survival. Included are numerous dichotomous keys to help in identifying some of the more common species of vascular plants, as well as how the species (or groups of species) were used by Native Peoples and early European settlers as medicine, food, and in crafts. Every attempt has been made to make the dichotomous keys simple, and technical terms have been used as sparingly as possible, and the descriptions include the most striking features that distinguish a given species.

Finally, it is impossible, upon completion of a book of this nature, to acknowledge appropriately the helpful assistance and inspiration given by numerous individuals who have aided in the conception and realization of what you currently hold in your hands. However, I would like to specifically recognize the following people and organizations for professional and personal reasons: Ms. Jennifer Reynolds (San Bernardino County Museum Association) and all the folks on the publishing committee that allowed this project to proceed; Dr. Timothy Krantz, University of Redlands; and the Lithuanian Scout Association Camp "Rambynas."

Ray Vizgirdas 2018
Boise, Idaho

TABLE OF CONTENTS

Disclaimer

Prefaces and Acknowledgements

INTRODUCTION

Ancient humans learned very quickly that they needed food to survive. Over the years (and generations) and through trial and error, humans developed a repertoire of edible, useful, and medicinal plants. It is because of these experiments that we have information on the uses of many plants. Unfortunately, today most people are unaware of these riches and have become dependent on only a few "domesticated" species of plants. Those that seek out wild plants find a continual source of stimulation and enjoyment, particularly as it relates to the survival of one's spirit.

The Mountains of Southern California

Living within the approximately 42,500 square mile confines of southern California are about 15 million people, or about 60% of the state's population. When one hears of southern California, most envision the cultural aspects and icons such as "LA", Hollywood, movie stars, the Valley, traffic, smog, and the numerous golf courses. To say the least, it is a busy and happening place. However, in light of all the glitter and fast-paced life portrayed, it is one of the few places in the United States where one can, in less than a day drive, go from watching whales and surfers at the beach near Santa Monica to pondering the mysteries of life in the alpine environment of Old Grayback (a.k.a. San Gorgonio).

Southern California is a region of great topographic and geological diversity. Included within its limits are high mountains ranges, valleys whose floors lie below sea level, precipitous canyons and broad basins at many elevations and an assemblage of other physiographic features that reflect a complex geologic history. Elevations vary from about 235 feet below sea level at the Salton Sea to 11,502 feet on Mt. San Gorgonio. Precipitation averages slightly over two inches a year at El Centro to more than 50 inches in the higher mountains. Temperatures vary from an extreme summer high of 125 degrees Fahrenheit at Indio to winter lows of -25 F in the higher mountains.

Given this setting, people in southern California recreate in many different ways. At one extreme, some stay within the confines of the city and enjoy what it has to offer. At the other end of the spectrum, there are those that must seek the refuge and solace of natural environments. There are those that head for the deserts to enjoy the heat and "primitiveness" of the land. While others head for the hills to enjoy the "high country." Those that head up to the local mountains go there to camp, hike, fish, hunt, ski, and just mess around. It is for those that head up to the mountains that this book was created.

The Transverse Ranges is made up of a series of east-west trending mountain ranges that include (from west to east) the Santa Ynez, Santa Monica, Liebre, Sawmill, Pellona, San Gabriel, San Bernardino, Little San Bernardino, Eagle, and Pinto mountains. Topographically, this Range embraces a rugged chunk of country featuring high, rough mountain masses and long, narrow, intervening valleys. Many of the higher peaks of southern California, outside the southern Sierra Nevada, are found in the Transverse range - Mount San Gorgonio at 11,502 feet in the San Bernardino Mountains is the highest mountain in southern California. Mount San Antonio (Baldy) in the San

Gabriel Mountains has an elevation of 10,064 feet and there are other peaks over 9,000 feet. The Transverse Range is nearly 300 miles long, extending from Point Arguello, 55 miles west of Santa Barbara, eastward to Eagle Mountains in the desert.

The Peninsular Ranges lies to the south of the Transverse Range and consists of a series of more/less north-south trending mountain ranges, most of them at moderate elevations. They include the rugged San Jacinto, Santa Rosa, Agua Tibia, and Laguna mountains. The Peninsular Ranges merge northward into the Los Angeles Basin, and the northwest geologic grain eventually terminates against the east-west Transverse Ranges. In gross aspect, the Peninsular Range is a large block uplifted abruptly along the eastern edge and tilted westward. The highest point, San Jacinto peak (10,831 feet) towers above Palm Springs and Coachella Valley. This scarp is as high as the east face of the Sierra Nevada and surpasses in vertical relief the east face of the Grand Tetons in Wyoming. The Peninsular Range actually belongs more to Mexico than the U.S. since they extend the entire 800 miles of peninsular Baja California.

The Forests and Vegetation

In order to better facilitate the learning of useful plants, it is helpful in knowing something about plant communities and plant distribution. For example, if you found yourself in a forest dominated by Ponderosa Pine (Pinus ponderosa) then you will also know that there are often a great many other plant species associated with this forest type. Each species has its specific distribution based on a variety of requirements.

One of the earliest attempts to explain the distribution of plants (and animals) was C. Hart Merriam, Chief of the U.S. Biological Survey. Here, he attempted to correlate climatic conditions, more specifically temperature, as the most important single factor in fixing the limits beyond which a particular species of plant or animal cannot go. Merriam's Life Zone System or Life Zones as it is commonly referred to, found great favor with biologists in the early years of the last century and was widely used. However, the limitations to this scheme were quite evident to everyone. It is not uncommon to still see some field guides today using the terminology developed by Merriam.

Today, we know that the patterns of plant distribution are controlled not only by changes in elevation, but by factors such as the availability of moisture during the growing season and the amount of snow accumulation during winter. Plant communities of warm, south-facing slopes can differ markedly from cooler north-facing slopes that lie just across a valley. Soil type also affects plants distribution. Soils that have developed on glacial till, as an example, often harbor different species than soils formed from decomposed granite.

In the late 1940's, two botanists, P.A. Munz and D.D. Keck, were in the process of developing a new flora for California, and felt a need for a more precise classification to describe the distribution of plants than that provided by Merriam. In 1949 they published "California Plant Communities", and divided the state into five biotic provinces that were further divided into 14 vegetation types. Within the 14 vegetation types they recognized 24 plant communities which includes Alpine and Montane Coniferous Forests.

Montane Forests

In southern California, these montane coniferous forests typically occur above 5,000 feet and are quite complex and change in composition and structure depending on precipitation, temperature, soils, wind, and frequency of fire. The classification of these forest communities in southern California is not easy as they are variable from site to site, but for our purposes they can be grouped or divided into four community types or zones. These include the lower montane zone ranging from 4,500 to 8,000 feet; the upper montane zone ranging from 8,000 to 9,000 feet; the subalpine zone from 9,400 feet; and the treeless alpine zone appearing at about 11,500 feet.

Lower Montane Zone

Coulter Pine Forest This species (*Pinus coulteri*) forms an open forest or woodland that is a transitional community. Coulter Pine ranges from approximately the San Francisco Bay Area down the western half of the state through the Coast, Transverse, and Peninsular ranges into Baja California. Because of this distribution, it is considered to be a Pacific species, not a Sierran. Individual trees in this forest are often widely spaced in this forest type that it can be called a woodland. In the areas of the San Gabriel, San Bernardino, and San Jacinto mountains where it is abundant, or the only pine, it forms the lowest zone of the montane coniferous forest above 3,000 feet. Generally, the understory includes species of the chaparral, shrubs, or other pineland annuals and perennial herbs. It also comprises a rather broad ecotone community between the mixed chaparral and the yellow pine forest.

Yellow Pine Forest This montane coniferous forest is dominated by two species of yellow pine - Ponderosa Pine (*Pinus ponderosa*) and Jeffrey Pine (*P. jeffreyi*), with Jeffrey Pine - being much more abundant in the southern California forests. In the Transverse and Peninsular ranges, Ponderosa Pine occupies the more mesic (moist) slopes on lower slopes than Jeffrey Pine. These two pines are closely related and often hybridize where they overlap. The open and park-like forests these species create are rather similar in appearance and have many associated species in common. But because they occupy different elevational and ecological zones, they are often treated as separate communities.

The Ponderosa Pine forest forms above the Coulter pine forest at elevations between 4,000 and 7,000 feet. Ponderosa Pine is rarely found on the transmontane (desert side) slopes of the Transverse Range because it is less tolerant of drought and low temperatures. It is also susceptible to smog damage. Fire suppression in Ponderosa Pine forests has led to its decline from dominance in many areas.

In contrast to Ponderosa Pine, the Jeffrey Pine forest appears to be more tolerant of serpentine soils, drought, low temperatures, deep snow, and smog, and equally tolerant of soil moisture and warm temperatures. In southern California, Jeffrey Pine dominates in the yellow pine forest at higher and more exposed sites, almost to exclusion of Ponderosa Pine. Elevation range is 6,400 to 8,300 feet.

Upper Montane Zone

The species usually associated with the Upper Montane Zone include: White Fir (*Abies concolor*), Mountain Juniper (*Juniperus occidentalis*), and Sugar Pine (*Pinus lambertiana*).

White Fir-Sugar Pine Forest While these species can be found on more mesic sites within the Yellow Pine forest, they also form their own community on moister, steep, north- and east-facing slopes and ridges at the higher elevations. This forest type can be found at elevations between 5,000 and 8,000 feet, and often include Incense Cedar (*Calocedrus decurrens*) as a component. White Fir forests are highly variable and are commonly dense with a rather lush appearance and multi-layered structure.

Mountain Juniper Woodland Stands of *Juniperus occidentalis* occupy dry, exposed ridges and slopes. While the Mountain Juniper is usually scattered and seldom dominant in the San Gabriel mountains, it is more abundant and reaches lower elevations in the San Bernardino mountains. This is essentially a high elevation tree that ranges down towards the deserts.

Subalpine Zone

Some subalpine coniferous forests can be found on the highest peaks of southern California, such as on Mount Pinos, and on the highest peaks of the San Gabriel, San Bernardino, San Jacinto, and Santa Rosa ranges. Two species of pine are usually associated with the subalpine zone - Lodgepole Pine (*Pinus contorta*) and Limber Pine (*P. flexilis*)

Lodgepole Pine Forest Above the Montane Coniferous forest the forest becomes even more open, and the trees becomes shorter in stature. The primary species that dominates above the Upper Montane zone is the Lodgepole Pine, a widespread species in western North America. In southern California, Lodgepole Pine is found in the San Gabriel, San Bernardino, and San Jacinto mountains where it forms extensive forests on the higher mountains from about 8,000 feet on the north slopes, and 9,000 feet on the south slopes to the tops of the mountains.

Limber Pine Forest Limber Pine dominates the upper subalpine forests of the desert-fronting mountains from Mount Pinos through the San Gabriel and San Bernardino mountains, to the San Jacinto and Santa Rosa mountains. Limber Pine is considered to be a poor competitor, and as a result it is usually restricted to eroded, step, rocky sites with well-drained, infertile soils that front desert regions and receive little precipitation. Like bristlecone pine, limber pine is a slow growing, long-lived tree. They are characterized by their soft, light, flexible wood that gives them their common name.

Alpine Zone

Only three of the highest peaks (over 11,000 feet) in southern California have Alpine communities - San Bernardino, San Gabriel, and San Jacinto mountains. These are treeless communities dominated by low-growing herbaceous plants and dwarf shrubs, but in comparison to the alpine environments in the Sierra Nevada, these alpine areas are floristically poor and fragmentary. They occur at elevations too cold to support trees.

AREA OF COVERAGE

The area encompassed by this book essentially is the Transverse and Peninsular Ranges in southern California. This area extends from Santa Barbara County south to San Diego County and includes major mountain areas and peaks such as:

- San Rafael, with Mount Pinos, Big Pine Mountain, Frazier Mountain, and Reyes Peak;
- Topatopa Mountains;
- Liebre and Sawmill Mountains;
- San Gabriel Mountains, with Gleason Mountain, Pacific Mountain, and Mt. San Antonio;
- San Bernardino Mountains with Mt. San Gorgonio;
- Santa Ana Mountains (a very small part);
- San Jacinto Mountains with San Jacinto Peak and Thomas Mountain;

- Santa Rosa Mountains including Toro Peak, Rabbit Peak, Hot Springs Mountain, and Mt. Palomar;
- Laguna Mountains with Cuyamaca Peak and Sheephead Mountain.

The general elevational extent is from about 4,000 feet on the ocean side (known as the cismontane) of the mountains and above 4,500 feet on the desert side (known as transmontane) of the mountains, up to over 10,000 feet.

SCIENTIFIC AND COMMON NAMES

In the discussion of edible and useful plants, plant families are arranged alphabetically within each of the four major groups of higher plants (ferns and their allies, gymnosperms, dicots, and monocots). While this arrangement may seem awkward to professional botanists, it has been adopted with the realization that this arrangement of families (and genera within families, species within genera) will be more easily consulted by readers in non-botanical fields who may have occasion to use this book.

The scientific and common names are given for each species. A brief description of the plant stressing key features is provided. Plant nomenclature (scientific and common names) used in this book follows that of "The Jepson Manual: Higher Plants of California (ed. J.C. Hickman 1993). Common names for plants can be misleading and do not always distinguish among the species. Additionally, a species known by a common name in one region may have another common name elsewhere, leading to further confusion. However, common names have been retained since they are generally of more interest, and more likely to be known, by the public. You are encouraged to learn to identify plants by both their scientific and common names.

Four basic categories of edible and useful mountain plants in southern California	
Category (plant group)	Approximate # of Species
Ferns and their Allies	> 22
Conifers	> 17
Flowering Plants – Dicots	> 585
Flowering Plants - Monocots	> 50
TOTAL	> 674

RARE AND PROTECTED PLANTS

The Native Americans were dependent upon nature for all of their needs and had an extensive knowledge of which plants (and animals) were edible or useful. Because of this dependence, they shared a strong conservation ethic based on the sanctity of life. Today, with increasing human population and our demands upon natural resources, many species are becoming rare and endangered due to habitat destruction, competition with non-native species, or other means.

Mountain ecosystems have evolved in the absence of human activities and have no ready response to some kinds of disturbance. Though appearing rugged, mountain environments are fragile, highly susceptible to disturbance, and generally have a low ability to rebound and heal after damage. The degree to which this is true is variable, but the vulnerability of mountain environments to disturbance is well documented (e.g., Price 1981, Zwinger and Willard 1972).

The flora and fauna of mountain ecosystems are composed of species that are well-adapted to cope with environmental extremes, low productivity, and fluctuations within the system. Because of climatic extremes, a brief growing season, lack of nutrients at higher elevations, low biological activity and productivity, their island-like character, steepness of slopes, and the basic conservatism of the dominant life forms all make for the rate of restoration to original conditions after disturbance in mountain environments rather slow (Price 1981).

Therefore, the plants you encounter on hikes and camping trips will vary in their ability to withstand harvesting. For example, collecting berries may not directly kill a plant, but may affect its ability to survive. Additionally, because digging up the roots of a plant will destroy it, you must select your specimens carefully. And remember, many species of wildlife rely on plants for survival, whereas most hikers and nature enthusiasts collect plants for pleasure, not necessity.

It is strongly recommended that you obtain a list of threatened, endangered or sensitive plant species before you start collecting. In California, the California Native Plant Society, California Department of Fish and Wildlife, and the U.S. Fish and Wildlife Service can be very helpful. By avoiding rare species and using common sense, you should be able to enjoy wild plants without appreciably affecting either their population or surroundings. You should also check in with the local land management agency (e.g., Bureau of Land Management or Forest Service) for their policy on collecting native plants.

Of recent years there has come into man's life a new joy. This joy is the acquaintanceship with plants. Nature has long been ready to reveal her secrets, but only to those prepared to hear and see. Gradually a new understanding has arisen between Nature and mankind, and as a result we obtain from such a revelation a joy undreamed of a few years ago.... Plants no longer are lifeless things labeled and grouped under ponderous Latin titles; they are highly developed organisms, which....walk, swim, run, fly, jump, skip, hop, roll, tumble, set traps, and catch fish; decorate themselves that they may attract attention; powder their faces; imitate birds, animals, serpents, stones; play hide and seek; blossom underground; protect their children, and send them forth into the world prepared to care for themselves.

- Royal A. Dixon

GUIDELINES FOR GATHERING

Since there are no general rules for distinguishing an edible plant from an unsavory or poisonous species, one must identify a plant correctly before attempting to use it. Some books even suggest that if you don't know a plant, you can eat a small quantity and wait to see if it has any adverse effects. This is a potentially serious mistake. For instance, if the unknown plant happens to be Death Camas (Zigadenus), not only would it cause much discomfort (such as a burning sensation in the mouth), it could kill you. Anyone who plans to search out and consume edible plants should exercise extreme caution. Correct identification of plants is necessary to avoid similar species or parts that may be unpalatable or poisonous. One of the best ways to learn about plants is to consult a knowledgeable botanist or qualified individual.

It is important to harvest plants with wisdom and respect. The uncontrolled harvesting of mountain plants could severely damage delicate plant communities. In addition, it may be illegal to injure or uproot a living plant in some areas covered by this handbook (e.g., National Monuments, State Parks). If a plant is rare or endangered, look for other edibles. If you are not in a survival situation, you should be even more frugal and thoughtful.

Also, be mindful of your own safety when gathering edible and useful plants. In California and elsewhere, County, State and Federal agencies often spray chemicals to control noxious weeds, especially in areas where logging, mining, and grazing activities occur, or in developed campgrounds and along roads. While such chemicals may be labelled as "safe," there are no guarantees. You should avoid collecting in areas affected by pollutants such as along roads or in drainages affected by mining activities.

In "*Stalking the Wild Asparagus*," Euell Gibbons described the taste for wild edible plants as an acquired one, ranging from awful to barely palatable. Although we have sampled many wild plants that fall into those two categories, we've also found wonderful delicacies that make supermarket food seem pale by comparison. If you take a positive outlook in your endeavor, it may someday help you if you were ever in a survival situation. Otherwise, it's a wonderful excuse to explore the western mountains.

"Where you find a people who believe that man and nature are indivisible, and that survival and health are contingent upon an understanding of nature and her processes, these societies will be very different from ours, as will their towns, cities and landscapes."

- Ian McHarg (1969)
"Design with Nature"

FERNS AND ALLIES

Fern and fern allies include the club-mosses, horsetails, and ferns. They are herbaceous plants that reproduce by spores, which develop inside structures called sporangia.

"Man has mental limitations, and nature is infinitely complex. To deal with this situation man invents classifications. Nature does not classify trees, flowers, and rocks - we do, so that we can deal with them in a reasonable fashion."

- R.P Sharp (1972)
"Geology Field Guide to Southern California"

WATER FERN FAMILY (Azollaceae)

These are small floating plants on water. Two species are known from California, but only the following species may be encountered in the southern California mountains at the lower elevations.

Mosquito Fern (*Azolla filiculoides*)

Description: This is a free-floating or stranded on mud aquatic fern that is triangular or polygonal in shape. It floats on the surface of the ponds, ditches, and other slow or sluggish waters. From a distance, it looks like a green-reddish carpet floating over the surface. It is common at elevations below 5,000 feet.

Ecology & Ethnobotany: Living within the cavities of this aquatic fern is a blue-green alga (*Anabaena azolla*). This is a nitrogen fixing alga that excretes nitrogenous compounds into the cavity from which *Azolla* can absorb them, making nitrogen available for both species. In addition to fixing nitrogen by utilizing light energy, the blue-green alga associated with *Azolla* also releases hydrogen from water. This is the first reported known photosynthetic system for producing hydrogen from water that is stable in air and requires only water as a hydrogen source. In nature the fixed nitrogen combines with the hydrogen to form ammonia that fertilizes the host *Azolla* plant. However, in the laboratory, the fern-alga relationship can be diverted from producing ammonia to producing only hydrogen gas. While these are small-scale experiments to date, there may be some promise in producing hydrogen on a larger scale via this biological method. Hydrogen yields more energy (on a weight basis) than any other non-nuclear fuel, which is as important finding in light of dwindling petroleum supplies.

Azolla has other important contributions to many developing countries. In areas such as Southeast Asia, Vietnam, and Singapore, it is used as a green manure. In these areas where rice is grown for human consumption, *Azolla* is deliberately cultivated. It is reported that in these rice fields, the yield of rice is 50 percent higher than normal. Pigs, cattle, and ducks are also fed *Azolla* in many parts of the world. In other areas, the plant is also useful in controlling mosquitoes and other weeds by blocking the water surface.

HORSETAIL FAMILY (Equisetaceae)

In this family, there is only one genus with about 15 species in worldwide. The plants have hollow, joined stems, and the leaves are whorled at the nodes, and are minute and toothed-like. None of the species are of economic importance.

Horsetail (*Equisetum*)

Description: In general, these are rhizomatous herbs with hollow, grooved, regularly jointed stems that are impregnated with silica. The leaves are reduced in size, appearing as a series of teeth around a joint. Spores are produced in cone-like structures atop the stems. They are found in moist soil along streams and rivers, marshes, and other damp habitats. The name comes from the Latin *equus*, for horse, and *seta* for bristles). Three species may be encountered in the southern California mountains.

Common Horsetail (*E. arvense*) This species occurs in damp or wet soils along streams or dryer sites following a disturbance.

Common Scouring Rush (*E. hyemale*) This species is found along streams, ditches, and ponds, often growing with *E. arvense* and seen from June to July.

Smooth Scouring Rush (*E. laevigatum*) This species occurs in wet or dry streamside soils and in marshes. Seen from June to July.

Ecology & Ethnobotany: Although all species are useful and identical in application, Common Horsetail is the most popular. The tough outer tissue can be peeled away and sweet inner pulp of all species can be eaten in small amounts. In large quantities, defined as greater than 20% of body weight by some authorities, they can be toxic. Certain chemicals in this plant are said to destroy specific B Vitamins such as thiamine. The enzyme thiaminase is apparently responsible for the poisoning. Cooking destroys this enzyme and renders the plants safe for consumption. The tuberous growth on the roots (actually rhizomes) can be eaten raw in the early spring or boiled later in the season.

Horsetails have an unusual chemistry. Some species contain alkaloids (including nicotine) and various minerals, whereas other species have been known to concentrate gold in their tissues, although not in

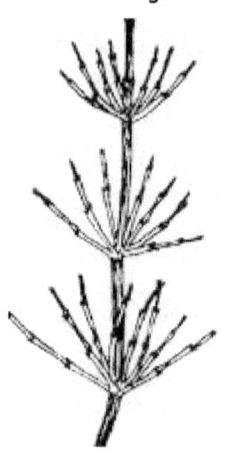

sufficient amounts to warrant extraction. In the fall, the stems become impregnated with silicon dioxide and can be used to scour pots and pans or as a type of sandpaper for wood. Many Native Americans used Horsetails to polish arrow shafts. The silica rich stems are reputed to be superior to the finest grades of steel wool. Additionally, the high silica content of this herb is said to be effective in strengthening bones and connective tissue.

Caution: The waters within which these plants grow may be contaminated.

QUILLWORT FAMILY (Isoetaceae)

These are small submerged or partially emersed plants with a corm-like stem crowned by numerous subulate or nearly filiform leaves. There are two genera and 75 species worldwide. None are of economic importance.

Quillwort (*Isoetes*)

Description: These are small, usually erect grass-like plants that are submerged or partially immersed in streams, ponds, and lakes. The stem, called a corm, is short, fleshy, 2 to 3 lobed, and has many roots developing from the base. The sporangia develop on the upper surface of the expanded leaf base. They are solitary, orbicular to ovoid, and usually covered by a thin membranous tissue called <u>velum</u>. The common name is derived from the plants' resemblance to a bunch of quills.

The approximately 150 species of Quillwort worldwide can only be distinguished by microscopic examination of their spores. Three species may be found in the southern California mountains.

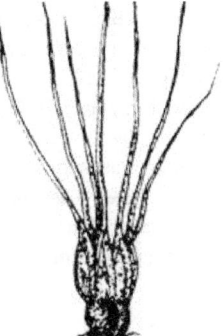

"Quick Key" to the Quillworts

1. Plants partially immersed or terrestrial - **2**
1. Plants mostly submerged and covered by water - *I. bolanderi*

2. Velum not well developed; leaves 2-11 inches long - *I. howellii*
2. Velum well developed; leaves $1\frac{1}{2}$ - 8 inches long - *I. nuttallii*

> **Bolander's Quillwort (*I. bolanderi*)** Look for this quillwort in shallow mountain lakes and ponds, from 5,000 to 11,500 feet.
> **Howell's Quillwort (*I. howellii*)** This species is found on the borders of lakes and ponds, usually below 9,000 feet.
> **Nuttall's Quillwort (*I. nuttallii*)** Nuttall's Quillwort is found in damp places, along banks of streams and rivers, below 9,000 feet.

Ecology & Ethnobotany: None of the species are known to be of economic importance. It is reported they have been occasionally used as food by people in Europe (Pheiffer 1922). We found no record of anything injurious about the plants. The corm serves as a storehouse of food for the plant, as do the subterranean winter organs of many perennials; thus, some food value is quite probable (Frye 1934). Weedon (1996) states that Quillworts are rich in starch and oils and could be edible raw or cooked. The corms are edible, but not palatable.

WATERCLOVER FAMILY (Marsileaceae)

These are aquatic or semiaquatic perennial herbs with creeping, hairy rhizomes that root in the mud. The leaves are 4-foliate and clover-like. The sporocarps are borne at the base of the stipe. The family is comprised of three genera and 70 species distributed worldwide. Two genera are native to the U.S. They are of limited importance as ornamentals.

Hairy Pepperwort (*Marsilea vestita*)

Description: This species is found in shallow lakes and ponds or on their muddy borders. It has also been found along slow-

moving streams. Arising from the rhizome are the distinctive leaves arranged like those of a four-leaf clover. The leaflets are pale green and hairy. Spores are produced within an elliptical, nut-like body borne on short stalks.

Ecology & Ethnobotany: While there are no documented uses for this species in southern California, related species are supposedly edible when dried and ground into flour. The sporocarps of Nardoo (*M. drummondii*) found in Central Australia can be crushed into a powder between stones, made into a dough by adding water, and baked into cakes.

ADDER'S-TONGUE FAMILY (Ophioglossaceae)

The herbaceous plants have fleshy rhizomes with numerous fibrous, often fleshy roots. The leaves (fronds) consist of two parts, a sterile simple or compound, sessile or stalked blade, and a stalked spore-bearing spike or panicle, sterile or fertile parts borne on an erect common stalk. There are three genera and 70+ species of this family. Two genera are found in the United States, but only Moonwort (*Botrychium*) may be found in the southern California mountains. None of the members of this family are of economic importance.

Moonwort (*Botrychium*)

Description: Worldwide, there are about 40 species of Moonwort, of which eight occur in California, and two in the southern California mountains. This is a difficult genus to study and requires fully developed specimens for identification. The spore sacs, technically termed the sporangia, are borne in grape-like clusters on a naked stalk, not on leaves as in the true ferns.

Ecology & Ethnobotany: Early records appear to indicate that juice extracts from Common Moonwort were used to stop bleeding and vomiting, and also for the treatment of bruises. They may have been used to concoct balsams for healing internal wounds (Grillos 1966).

Foster and Duke (1990) describe a root poultice or lotion made from *B. virginianum* (Rattlesnake Fern) that was used for snakebites, bruises, cuts, or sores. In folk medicine, the root tea was used an emetic.

FERN FAMILY ("Polypodiaceae")

Botanists have reclassified many fern taxa into other families including Dennstaediaceae, Dryopteridaceae, Polypodiaceae, and Pteridaceae). For convenience, and until we are able to completely revise and update this handbook, they are treated here as a single family – "Polypodiaceae." Generally speaking, fern fronds can be used as makeshift place mats and interspersed between layers of food in cooking pits.

Caution: A number of survival handbooks (e.g., Risk 1983) suggest that most ferns in their young stage are edible. However, in most cases, they are referring to Bracken Fern (*Pteridium*). It is best to avoid ferns unless you are able to positively identify them. Many plants such as Poison Hemlock (*Conium maculatum*) have delicately cut leaves, which to the untrained eye could easily be mistaken for a fern.

"Quick Key" to the Ferns

1. Fronds of two types; sterile fronds with broad lobes or leaflets - *Cryptogramma*
1. Fronds all similar - **2**

2. Sori located at the edge of leaflets and covered by curled-under margins - **3**

2. Margins of leaflets never curled under and covering the sori - **6**

3. Fronds 2-4 feet long - *Pteridium*

3. Fronds shorter - **4**

4. Blades densely scaly or hairy beneath - *Cheilanthes*

4. Blades glabrous or nearly so beneath - **5**

5. Margins partly curled under; leaflets on stalks - *Adiantum*

5. Margins completely curled under; leaflets sessile - *Pellaea*

6. Fronds once pinnate; leaflets entire, dentate, or occasionally lobed at base – **7**

6. Fronds 2 to 3 pinnate, or if one pinnate the leaflets deeply lobed - **8**

7. Frond pinnately lobed, the lobes attached to midrib by a broad base - *Polypodium*

7. Leaflets distinct, attached to midrib by a narrow stalk - *Polystichum*

8. Sori covered by a curved indusium; fronds more than 15 inches long - **9**

8. Sori not covered or indusium scale-like - **10**

9. Indusium slightly curved, elongate - *Athyrium*

9. Indusium horseshoe-shaped, circular in outline - *Dryopteris*

10. Indusium present and covering sori; petioles smooth - *Cystopteris*
10. No indusium or inconspicuous; petioles scale at base - *Woodsia*

Maidenhair Fern (*Adiantum aleuticum*)

Description: There are about 200 species of *Adiantum* that occur worldwide. The underground stems of this species are short, thick, and have brownish scales. The leaves close, erect 12 to 30 inches long. The blades are divided into 2 equal parts, each curving part with 3 to 8 pinnae. The leafstalks are dark and scaly at the base. The pinnae are long-oblong 3 to 9 inches long. This species occurs in moist shaded rock crevices and canyons from sea level up to 11,000 feet.

Ecology & Ethnobotany: These black-stemmed ferns can be used in basketry. The herbage is reported to be bitter and causes an increased secretion of mucus. The leaves were used as a tea or syrup to treat colds, coughs, and hoarseness. Rhizomes were used as a stimulant, to soothe the mucous membranes of the throat and loosen phlegm. Some of these medicinal uses have been recorded since Classical times.

Ladyfern (*Athyrium filix-femina*)

Description: Ladyfern is a medium-sized plant growing in moist, shady places from 3,500 to 8,000 feet.

The leaves are 1-3 pinnate and clustered. The sori are oblong to horseshoe shaped. The indusium is thin and fragile, soon disappearing or sometimes lacking. The name is from the Greek *a*, meaning without, and *thurium*, referring to a long oblong shield. This species is widely distributed in North and South America.

Ecology & Ethnobotany: Ladyfern fiddleheads were eaten in the early spring when they were 2 to 6 inches tall. They were boiled, baked or eaten raw with grease. Grillos (1967) further indicates that Ladyfern has certain chemical properties for medicinal use. The underground stems, pulverized to a powder have been used to drive worms out of the intestinal system, although this use has not been medically recognized. A tea made from the root was used as a diuretic, and the powdered root was used externally for sores.

Coville's Bead Fern (*Cheilanthes covillei*)

Description: The fronds of this fern are ovate in outline, and the pinnae are densely hairy below the margins. There are many leaves, 4 to 12 inches long, ovate to ovate-lanceolate, with brown to dark-purple stalks. It is common in rocky places between 1,500 and 9,000 feet. The genus name is from the Greek, margin and flower, referring to the color of the stipes.

Ecology & Ethnobotany: This species was used medically by Hupa women during childbirth. The Kawaiisu drank a tea made from the stems and leaves as a general tonic.

Rock-brake (*Cryptogramma acrostichoides*)

Description: Rock-brakes are small ferns of rocky places, mostly at high elevations. These plants grow in dense tufts up to 16 inches tall. The sterile leaves are dark green, tri-pinnate, and the stipes yellow or straw-colored to the base. Fertile leaves are larger and longer. This species occurs on cliffs, ledges, or talus slopes from 6,000 to 11,000 feet. The genus name is from the Greek cryptos, meaning hidden, and gramme, meaning line, and refers to the sori being covered by the infolded margins of the pinnules

Ecology & Ethnobotany: A related species (*C. sitchensis*) was used as an eyewash. Here they made an infusion of the washed, strained fronds. The infusion was also taken gallstones.

Fragile Fern (*Cystopteris fragilis*)

Description: This 6 to 16-inch tall plant is loosely tufted from a short creeping rhizome. The leaves are thin and delicate in texture. Stipes are brown below, yellowish above, and smooth. The indusia are small, attached at one side and arching back to form a hood. This is a widely distributed fern, and is found in the crevices of cliffs and ledges, in soil under rocks, shrubs, or trees, between 5,000 and 9,000 feet.

Ecology & Ethnobotany: This plant was used as a dermatological aid by the Navajo. Here, a cold, compound

infusion of the plant was made and used as a lotion for injuries.

Woodfern (*Dryopteris filix-mas*)

Description: This is a medium sized fern arising from rhizomes. The leaves are bi- to tri-pinnate. The sori are round in outline, occurring on the veins, in two rows to each pinnule. The indusium is attached at one side by a deep notch. This species occurs in cool shaded ravines, often in cliffs or talus between 5,500 to 10,000 feet. Another locally common species is *D. arguta*. It can be identified by examining the leaflets. In short, the longest primary leaflets are near the base. The name is from the Greek *drys*, meaning oak, and *pteris*, meaning fern.

Ecology & Ethnobotany: The edibility of the various species of Woodfern is unknown, but some species are reported to be edible, poisonous, or of medicinal value. Several species contain phloroglucinol derivatives ("filicin"), which paralyze intestinal parasites. For example, a root tea from *D. cristata* (Crested Woodfern) was traditionally used to induce sweating, clear chest congestion, and expel intestinal worms (Foster and Duke 1990). In addition, an oleoresin was extracted from the roots of *D. filix-mas* (Male Fern) to expel worms.

Cliff-brake (*Pellaea*)

Description: These are small tufted ferns growing in crevices of rocks. The rhizomes are short, thick, creeping and densely brown scaly, and covered with old stipes. The leaves are singly pinnate or bi-pinnate. The stipes are dark reddish-brown. The sori are covered by the recurved margin of the leaf segments. The generic name comes from the Greek *pellos* (dusky), and refers to the appearance of the leafstalks. Two species can be found in the southern California mountains.

"Quick Key" to the Cliff-brakes

1. Blades once pinnate - *P. breweri*
1. Blades 2 to 4 times pinnate - *P. mucronata*

Brewer's Cliff-brake (*P. breweri*) This species is found on exposed rocky slopes between 7,000 and 11,000 feet.

Bird's-foot Cliff-brake (*P. mucronata*) This species can be found in exposed or partially shady, rocky places mainly below 6,000 feet.

Ecology & Ethnobotany: A refreshing tea was said to be made by the Luiseno Indians from *P. mucronata*. They also used the tea medicinally as a decoction to stop hemorrhages, to reduce fevers, and as an emetic. The decoction was used as a wash for skin problems. Other Native Americans used the brown fibers from the rhizome to make basketry patterns.

Caution: The younger leaves and stems are occasionally eaten by sheep and other grazing animals. However, they are poisonous and frequently cause death when eaten.

Western Polypody (*Polypodium hesperium*)

Description: This is a shade-loving fern. The fronds of this fern are pinnatifid to nearly pinnately compound and attached to creeping rhizomes by a distinct articulation. The sori are round in outline, and there is no indusium. This species is reported to occur in the San Bernardino and San Jacinto mountains, growing on rock ledges and crevices between 5,000 and 8,500 feet. The genus name is Greek, meaning many and foot, alluding to the numerous protuberances on the rhizomes.

Ecology & Ethnobotany: The rhizomes were chewed or an infusion of the rhizomes was made for colds and sore throats. The rhizomes were also used as medicine for sore gums. The rhizome has a pleasant, sweet taste, almost like licorice. The roots of related species (*P. californicum* and *P. glycyrrhiza*) were also eaten by California Natives. Other Natives Americans used the bruised roots to heal sores and to relieve rheumatic pain. An extract of the root of *P. californicum* was used to treat internal injuries and kidney ailments.

Eaton's Shield Fern (*Polystichum scopulinum*)

Description: These ferns are small to medium sized that arise from short, stout rhizomes with light brown scales. The leaves are coarse, evergreen, pinnate, oblong to lanceolate in outline. Pinnae are more/less deltoid, lobed or divided near the base. This species is found on dry cliffs and rocky places from 5,000 to 10,500 feet. The name is from the Greek *polys* (many), and *stichos* (row), because the sori of some species develop in several rows.

Ecology & Ethnobotany: In general, the leaves of these ferns can be used as a protective layer in pit cooking, as flooring or bedding. While the edibility of this species is unknown, the large rhizomes of a related species, *P. imbricans* (Swordfern), can be roasted over a fire or steamed in a pit oven, then peeled and eaten. The cooked rhizomes are also said to be a cure for diarrhea.

Western Brackenfern (*Pteridium aquilinum*)

Description: This is a medium- to large-sized plant with decompound, broadly triangular leaves up to 7 feet long including the stipe. The stipes are green or yellowish, and there are fine white hairs. The sori are marginal and continuous, and partially covered by the recurved leaf margins. This is a widely distributed species found in open woods, rock slides or slopes in damp or dry places, up into the high mountains. The generic name comes from the Greek pteris, a wing, and is applied to ferns because of their feathery leaves.

Ecology & Ethnobotany: The young fronds, or fiddleheads of Bracken Fern can be collected, boiled and dried in the sun. The dried product can then be used as a winter food. Old fronds may be poisonous in large amounts (see Caution below). The starchy rhizome (underground stem) is edible after roasting or boiling, but is usually tough. The leaves can be used as one of the protective plant layers for pit cooking. Some Native Americans would consume only fiddleheads so that their scent would not scare off deer. A root tea was used for stomach cramps and diarrhea, and poulticed roots were used for burns and sores. Ashes of the plants have been used as an ingredient to make glass and soap.

Caution: While this plant has traditionally been accepted and harvested as a suitable edible, there is new evidence indicates that eating sufficient quantities over a period of time may be dangerous to your health (Foster and Duke 1990). The plant is known to contain several poisonous compounds, including a cyanide-producing glycoside (prunasin), an enzyme, thiaminase, that reduces the body's thiamine reserves, and at least two potent carcinogens, quercetin and kaempferol. Another, unidentified toxin is believed to be naturally occurring, radiation-mimicking substance, also apparently mutagenic and carcinogenic. Bracken has caused many livestock deaths. The risks to humans of eating Bracken fiddleheads and rhizomes have not been fully established, but their safety is questionable. Schofield (1989) reports that it is currently suspected of causing stomach cancer in Japan.

Woodsia (*Woodsia*)

Description: These small ferns commonly grow in rocky places. The underground stem is densely tufted and clothed with broad, thin scales. The leaves are clustered, numerous, small, linear to lanceolate-ovate, and once- or twice-pinnate. The sori are round and seated on the back of the free veins, and the indusia is under the sori with star-shaped divisions. The genus name honors Joseph Woods, an English botanist. Two species may be encountered in the southern California mountains.

"Quick Key" to the Woodsias

1. Leaves smooth - *W. oregana*
1. Leaves glandular - *W. scopulina*

Oregon Woodsia (*W. oregana*) This species occurs in dry rocky places from 4,000 to 11,000 feet.

Rocky Mountain Woodsia (*W. scopulina*)
Similar to Oregon Woodsia, but the blades are finely glandular to distinctly pubescent. This species occurs on exposed rocks from 4,000 to 11,000 feet.

Ecology & Ethnobotany: Rocky Mountain Woodsia was used as a sign of water when traveling through the mountains by the Natives further to the north.

SPIKEMOSS FAMILY (Selaginellaceae)

This is a family of moss-like plants that includes many important natural ground covers on rocky or gravelly soils. The sporangia are borne in terminal spikes of sporophylls, the larger ones containing 3 or 4 megaspores, the smaller ones numerous, minutes microspores. Members are sometimes called "little clubmoss" because of their resemblance to the Clubmoss Family (Lycopodiaceae).

Bluish-spikemoss (*Selaginella asprella*)

Description: In general, Spikemosses are low, moss-like, leafy, evergreen terrestrial plants, and their leaves are scale-like less than 1/8-inch long. Bluish-spikemoss is one species that may be encountered in the San Gabriel, San Bernardino, and San Jacinto mountains. It may be found in a variety of habitats including dry rocks, ledges, cliffs, crevices, talus slopes, meadows, and lake shores up to timberline.

Ecology & Ethnobotany: The spores of bluish-spikemoss were once used like that of Clubmoss

(*Lycopodium*) in making pills not adhere to one another. Many campers still gather the plants with mosses and other plants to make a soft bed when sleeping on the ground.

While there are no edible or medicinal uses recorded for southern California mountain species, related species, such as *S. concinna* and *S. obtusa* found on the Reunion Islands in the Indian Ocean were used medicinally as astringents, blood purifiers, and as carminatives in cases of dysentery. In Mexico, a variety of *S. rupestris* was used as a home remedy, and in the East Indies, *S. convoluta* was considered to be an aphrodisiac.

GYMNOSPERMS

Gymnosperms are a group of plants that do not have flowers and fruits. The seeds are exposed, usually on a cone-like structure. Hence, the majority of the gymnosperms are cone-bearing, most commonly known as conifers: spruce, fir, pines, and junipers.

The only gymnosperm that does not resemble other gymnosperms is an ancient plant called Ephedra or Mormon Tea. Ephedra is a desert shrub with broom-like branches that appear leafless. Leaves are minute and scale-like, and opposite or whorled at the nodes. At flowering time in the spring, small cones from at the nodes.

"Quick Key" to the Gymnosperms

1. Leaves needle-like (Pinaceae) - **2**
1. Leaves minute, broom-like or scale-like - (Ephedraceae, Cupressaceae) - **12**

2. Needles 2 to 5 in a cluster - **3**
2. Needles occurring singly - **10**

3. Needles 5 in a cluster - **4**
3. Needles less than 5 in a cluster - **5**

4. Needles long and slender, 3-4 inches long; cones 10-18 inches long - **Sugar Pine (*Pinus lambertiana*)**
4. Needles stout and stiff, $1\frac{1}{2}$-3 inches long; cones 3-10 inches long - **Limber Pine (*P. flexilis*)**

---

5. Needles 4 in a cluster - **Parry Pine** (*P. quadrifolia*)
5. Needles less than 4 in a cluster - **6**

6. Needles 2 in a cluster - **Lodgepole Pine** (*P. contorta*)
6. Needles 3 in a cluster - **7**

7. Cones 3-6 inches long - **Ponderosa Pine** (*P. ponderosa*)
7. Cones 6-12 inches long - **8**

8. Cone scales not ending in recurved hooks; montane tree
- **Jeffrey Pine** (*P. jeffreyi*)
8. Cone scales ending in recurved hooks; low elevation tree
- **9**

9. Leaves drooping; cones 6-10 inches long - **Gray Pine** (*P. sabiniana*)
9. Leaves erect; cones 9-14 inches long - **Coulter Pine** (*P. coulteri*)

10. Cones erect; scales closely imbricated, deciduous; needles blunt tipped - **White Fir** (*Abies concolor*)
10. Cones pendant; scales not as above; needles with sharp points - **11**

11. Cone scales with overlapping bracts; cones 4-6 inches long - **Big Cone Spruce** (*Pseudotsga macrocarpa*)
11. Cone scales without overlapping bracts; cones squat, $1\frac{1}{2}$-$2\frac{1}{2}$ inches long - **One-leaf Pinyon Pine** (*Pinus monophylla*)

12. Branches broom-like and appearing leafless - **Ephedra (*Ephedra viridis*)**
12. Branches with numerous overlapping, green, scale-like leaves - **13**

13. Fruit a blue-black "berry"; tree or shrub - **Junipers (*Juniperus*)**
13. Fruit a woody cone - **14**

14. Cone globose - **Tecate Cypress (*Cupressus forbesii*)**
14. Cone not globose; scales open out - **Incense Cedar (*Calocedrus decurrens*)**

EPHEDRA FAMILY (Ephedraceae)

These are shrubs or small trees with opposite or whorled, jointed branches. Leaves are foliaceous or scale-like. There is one genus with about 40 species that are found in warm arid regions of the Old World and New World. An Asiatic species is the source of ephedrine.

Ephedra (*Ephedra viridis*)

Description: This is an erect shrub with broom-like, leafless, yellow-green branches. The leaves are scale-like at the joints and are seldom noticed. Leaf scales are opposite, about 1/8-inch long, and deciduous. Cones are about $\frac{3}{4}$-inch long that occur 2 or more at the nodes. Ephedra is common on dry rocky slopes from 3,000 to 7,000 feet.

Ecology & Ethnobotany: The name "Ephedra" is of ancient times, as the plant has been in human history for thousands of years. Long ago the Chinese realized the medicinal properties of various species of Ephedra for treating respiratory ailments, and we now know that Ephedra taken orally does stimulate the body in a manner similar to injected adrenaline. Ephedra is now synthetically produced under the name of ephedrine and is one of the leading over-the-counter and prescription treatments for allergies, congestion, and asthma. Our common species, *Ephedra viridis*, has little or no ephedrine in it. Western U.S. residents have for many years brewed what some consider a soothing tea from stems. "*Viridis*" is Latin for "green," referring to the green stems.

Please Note: The stems may contain very small amounts of caffeine and ephedrine (a drug that acts like adrenalin and epinephrin). The closely related pseudoephedrine is now synthesized commercially and is an ingredient in commercial asthma and cold remedies. Pseudoephedrine is also a precursor in the production of the dangerous illegal drug methamphetamine ("speed"). A tea with stimulant properties is made by steeping dried stems. It has been used medicinally to treat a variety of ailments including syphilis, diabetes, and pneumonia. A Chinese species is the source of *ma huang*, a tea so potent that it has caused deaths from overstimulation of the heart.

CYPRESS FAMILY (Cupressaceae)

There are 19 genera and 130 species in this family. Five of the genera are native to the United States. The family is economically important as timber trees and ornamentals.

"Quick Key" to the Cypress Family

1. Fruit a blue-black fleshy cone or "berry"- **Junipers** (*Juniperus*)
1. Fruit a woody cone - **2**

2. Cone globose - **Tecate Cypress** (*Cupressus forbesii*)
2. Cone not globose; scales open out - **Incense Cedar** (*Calocedrus decurrens*)

Incense-cedar (*Calocedrus decurrens*)

Description: This is a tall tree that grows 75 to 125 feet tall. It has evergreen, aromatic herbage and red-brown shedding bark. The branches form graceful, flat sprays. The leaves are scale-like and closely appressed to the branches. The cone is small, about 1 inch long, and appears almost bird-like with 6 scales. Incense-cedar is fairly common on the higher mountain slopes, particularly where there is sufficient moisture. In most places in southern California it is found above 6,000 feet.

Ecology & Ethnobotany: This species was formerly assigned to the genus *Libocedrus*. This plant was used as a drug, for food and fiber. As a food, the leaflets were added as flavoring to while leaching acorn meal. As a drug, the branches and twigs were used in sweat bathes. The Paiute's made an infusion of the leaves and the steam was inhaled for colds, whereas other Natives made a decoction of the leaves for stomach troubles.

The wood of Incense Cedar can be used in fire-by-friction fire starting methods. The bark and wood can be used in weaving baskets. The roots were used as warps and weft in baskets too.

Tecate Cypress (*Cupressus forbesii*)

Description: This evergreen tree grows less than 30 feet tall, and has red-brown, smooth bark that peels. The leaves are scale-like and closely appressed to the stem. Cones are globose, 1 inch or more in diameter, and without projecting cones scales. Tecate Cypress occurs at lower elevations on dry slopes in Orange and San Diego counties.

A number of different trees around the world go by the name cypress. However, only those in the genus *Cupressus* are true cypresses. They look very much like Junipers (*Juniperus*), but the tiny, twig-hugging leaves are not in flat sprays.

Ecology & Ethnobotany: Decoction of stems taken for rheumatism and colds. Berries ground into meal and used with bread dough. The berries were eaten raw or roasted.

Junipers (*Juniperus*)

Description: These are evergreen trees and shrubs, with opposite or whorled, scale-like or linear leaves. The male cones are small, and the female cones are larger and "berry-like." Three species can be found in the southern California mountains.

"Quick Key" to the Junipers

1. Tree; mature berries red or reddish brown and $\frac{1}{4}$-inch long; higher montane areas - *J. occidentalis*
1. Tree or shrub; mature berries blue or blue-black and $\frac{1}{2}$ to $\frac{3}{4}$-inch long; lower desert slopes - **2**

2. Leaf gland obvious; plant appearing shrub-like - *J. californica*
2. Leaf gland obscure, usually tree-like on desert side of mountains - *J. osteosperma*

California Juniper (*J. californica*) This shrub has no distinct trunk below the branches. California Juniper is common on desert mountains and dry slopes, particularly associated with pinyon pine or Joshua tree. Found at elevations below 5,000 feet.

Western Juniper (*J. occidentalis*) This is a low tree 15 to 60 feet tall with a well-defined trunk. Western Juniper occurs on dry mountain slopes in the San Bernardino mountains and northward, in most places at the higher elevations.

Utah Juniper (*J. osteosperma*) This tree has thin, gray-brown bark that ages ash-white and is found from 4,000 to 8,000 feet on the desert side of the San Gabriel and San Bernardino mountains.

Ecology & Ethnobotany: Junipers offer countless products. The berries and twigs can be used to make tea, season game, smoke fish, repel moths, soothe rheumatic pains, and kill infectious germs. The fleshy cones are edible raw, but taste better if dried, ground, and used as a flour, flour extender, or made into cakes.

Cooking the flour with other foods can make it more palatable. The berries can also be roasted and ground, and used as a coffee substitute. The Swedes make an extract from the berries, which they generally eat with bread, in much the same way we use butter. In addition, the berries have been used to give gin its characteristic flavor.

The boughs can be steeped in hot water for 5-10 minutes to make a beverage. Cooking them in an uncovered pot is recommended to allow the volatile oils to escape. A leaf or berry infusion was used to relieve urinary problems.

The shredded bark is an excellent tinder for primitive fire-starting techniques and can be used as bedding and padding. It is said that some Native American tribes ate the inner bark in times of famine. The inner bark was also used for clothes and mattresses and could be worked with the hands until soft enough to use for baby diapers or sanitary pads.

Juniper oil extract can be rubbed on skin as an insect repellent, and to relieve pain in muscles and joints. The bark, roots, twigs, and cones furnish red dyes. The white film covering the berries is a type of yeast that can be used to make a primitive sourdough starter.

Ash Cakes

Make a dough by mixing flour with water. Pat the dough into a patty. The thicker it is, the more doughy it will be, whereas the thinner it is the crispier it will be. Throw the patty into the ashes (hot coals), let it cook a bit, and you have an ash cake.

PINE FAMILY (Pinaceae)

There are 10 genera and 250 species in this family worldwide. Six genera are native to the United States. The pine family is comprised of trees with needle-like leaves. Except for Larches (*Larix*), most species are evergreen. Many are of economic important timber trees.

"Quick Key" to the Pine Family

1. Needles occurring in clusters of 1-5 with a papery bundle sheath at the bases of them - **Pines (*Pinus*)**
1. Needles occurring singly, no papery bundle sheath at the base - **2**

2. Leaves arise from round, flat scars; seed cones are erect and have deciduous scales, so cones are rarely found intact on the forest floor - **White Fir (*Abies concolor*)**
2. Leaves arise from pegs; seed cones are pendant at maturity and have persistent scales, so cones drop intact to the ground - **Big Cone Spruce (*Pseudotsuga macrocarpa*)**

White Fir (*Abies concolor*)

Description: This tree grows up to 200 feet tall and exhibits a pyramidal growth pattern. Leaves are needle-like and spirally arranged on the branches and have blunt tips. The cones are erect and found near at the top of the tree. Cones are oblong shaped and the scales are closely imbricated and fall off separately when they open to release the seeds. This is why it is very uncommon to find intact fir cones around the base of the tree. White Fir grows on moist mountain slopes at slightly higher elevations than the first occurrence of Jeffrey pines. It is common in southern California mountains from 6,000 to 9,000 feet.

Ecology & Ethnobotany: Native Americans used the needles for tea. Medicinally, an infusion of the foliage taken and used as a bath by the Acoma and Laguna Indians for rheumatism. Some Native Peoples in Nevada used a decoction of White Fir bark resin and needles to help pulmonary troubles. A simple poultice or warm pitch of resinous sap from the bark or large branches could be applied to sores or boils. The Tewa Indians used the sap from the main stem and larger branches for cuts. A decoction of the resin can be taken for venereal diseases. The resin has been known to be used by early New Mexico natives to fill decayed teeth. Extracts from the bark

have shown anti-tumor activity. One of the active materials might be a complex tannin.

Historically white fir was considered undesirable for timber. Now that the availability of premium timber species has declined, white fir is being recognized as a highly productive and valuable tree species and is widely used in the wood products industry. White fir is a general, all-purpose, construction-grade wood used extensively for solid construction framing and plywood, and to a lesser extent, for pulpwood. It is also used for poles and pilings because of its straight grain and low taper, but requires large amounts of preservatives because the heartwood decays rapidly. It is poorly suited for firewood because of its low specific gravity and heat production (80% as much heat by volume as Douglas-fir produces), but it is used for firewood anyway.

"One does not usually think of conifer trees and forests as sources of food. To Native Californians of the past, however, these forests were great orchards. While oak acorns symbolically provided bread for subsistence, pine nuts provided the cake."

- Barbour et al. (1993)
California's Changing Landscape

Pines (*Pinus*)

Description: Pines may be divided into two major subgroups - "soft" pines and "hard" pines. In soft pines, the needles are usually in bundles of five and the cones have no prickles. The wood of the soft pines is straight grained, that is comparatively free from resin, and easy to work. It is used for rough carpentry, cabinetwork, patterns, toys, crates, and boxes. In contrast, hard pines have 2-3 needles per bundle, and the cones have prickles. The strong, resinous wood of hard pines is used in buildings, bridges, ships, and other types of heavy construction. Because of its durability, the wood of hard pines is valuable for floors, stairs, planks, and beams. Nine species of pines can be found in the southern California mountains. Because pines are a relatively easy group of plants to identify and are safe in terms of edibility, we have provided a dichotomous key to the species below.

"Quick Key" To *Pinus*

1. Needles occurring singly - *P. **monophylla***
1. Needles 2 - 5 in a cluster – **2**

2. Needles 5 in a cluster - **3**
2. Needles less than 5 in a cluster - **4**

3. Needles long and slender, 3-4 inches long; cones 10 to 18 inches long - *P. **lambertiana***
3. Needles stout and stiff, $1\frac{1}{2}$ - 3 inches long; cones 3 - 10 inches long - *P. **flexilis***

---

4. Needles 4 in a cluster - *P. quadrifolia*
4. Needles less than 4 in a cluster - **5**

5. Needles 2 in a cluster - *P. contorta*
5. Needles 3 in a cluster - **6**

6. Cones 3 - 6 inches long - *P. ponderosa*
6. Cones 6 - 12 inches long - **7**

7. Cone scales not ending in recurved hooks; montane tree - *P. jeffreyi*
7. Cone scales ending in recurved hooks; low elevation tree - **8**

8. Leaves drooping; cones 6 - 10 inches long - *P. sabiniana*
8. Leaves erect; cones 9 - 14 inches long - *P. coulteri*

    Lodgepole Pine (*P. contorta*) Lodgepole Pine is common at high elevations, in some places forming pure stands.

    Coulter Pine (*P. coulteri*) Coulter Pine occurs on dry slopes at the lower elevations, usually below the Yellow Pine belt.

    Limber Pine (*P. flexilis*) Limber Pine occurs on dry mountain slopes, from 7,500 to 11,000 feet.

    Jeffrey Pine (*P. jeffreyi*) The bark has the aroma of vanilla or butterscotch. The scale tips have a sharp, prickly point that turns down, not out as in Ponderosa Pine (hence the saying, "prickly Ponderosa, gentle Jeffrey"). Jeffrey Pine is quite common in the southern California mountains.

Sugar Pine (*P. lambertiana*) Young cones are coated with a sugary sap that looks almost like ice. Sugar Pine is common in the upper montane zone, from 6,000 to 10,000 feet and first appears in most places after Jeffrey Pine is well established.

Pinyon Pine (*P. monophylla*) One-leaf Pinyon Pine is common on desert mountain slopes between 3,500 to 5,500 feet.

Ponderosa Pine (*P. ponderosa*) The cones scales have a short prickle at the tip that points outward. Ponderosa Pine is found from 5,000 to 9,000 feet.

Parry Pine (*P. quadrifolia*) Parry Pine grows on desert slopes at the lower elevations (below 5,500 feet).

Gray Pine (*P. sabiniana*) Gray Pine grows on dry slopes and ridges below 5,000 feet. It occurs from northern Los Angeles county north, but is not common in southern California. This species was once referred to as Digger Pine. Since this name is degrading to the Native Americans, it is best to avoid its usage.

Ecology & Ethnobotany: All pines have edible seeds. However, they are an erratic food source, yielding an abundant crop in some years and a sparse crop in others. To collect the cones, long poles were used to knock them from the branches. One of the best ways to gather seeds is to heat the green cones until they open. The seeds are best when harvested in fall or early winter when cones normally release their seeds. The nutritious seeds can then be shelled and eaten, or ground or roasted and made into a flour. Seeds may contain as much as 15%

protein, 50-62% fat, and 18% carbohydrates, with approximately 3,000 calories per pound.

The inner bark is also edible in an emergency. Though tedious, the tender mucilaginous layer between the bark and wood was scraped or peeled off. It was then cooked or ground into meal. The use of the inner bark by Native Americans was so extensive that early explorers reported recorded large areas of trees stripped of bark.

The firm and unexpanded pollen cones can be boiled and eaten. They have a surprisingly sweet and non-pitchy taste. The edible pollen is usually mixed with flour and used as a soup thickener.

"Pine seeds have outstanding food value, and are especially high in fat content. Modern diets tend to focus on protein and to play down the fat, but fat is highly important to people exposed to low temperatures and lacking warm clothing, as was the case among aboriginal Californians. The fifty percent fat content may have been a more important survival factor than the twenty-five percent protein content."

- Barbour et al. (1993)
"California's Changing Landscape"

The needles of most pines can be steeped in hot water to make a satisfying tea and are a good source of Vitamin C. It also takes some practice to steep the right amount of leaves, since too much may be too strong. Additionally, the pine cleaning fluid can be extracted from boiling the needles and skimming off the oil-like substance from the surface. It may take a lot of pine needles to get a small cupful.

Pine sap can be collected in quantity from cuts, burns, and broken branches. The collected sap is then heated and formed into balls for future use. Be careful not to expose the sap to flames as it is very flammable.

All species of pine have been used medicinally for many centuries. Chewing the sap was said to be soothing for a sore throat. The sap can be dried, powdered, and applied to the throat with a swab. It was also heated and used as a dressing to draw out embedded splinters or to bring boils to a head. Smeared on a hot cloth the sap was used much like a mustard plaster in treating pneumonia, sciatic pains, and general muscular soreness. Pine oil is widely used in massage oil for muscular stiffness, sciatica, rheumatism, and in vapor rubs for bronchial congestion. All pines are rich in resin and camphoraceous volatile oils, including pinene, that are strongly antiseptic and stimulant. The needles and resin produce a brown dye.

Pine roots were valued as twining material for baskets. The roots, about as thick as a pencil, can be several feet long. Roots were collected from opposite sides of a tree each year to prevent destroying the tree. After cleaning, roots were buried in a heated pit. Fire built over the pit for two days turned the roots a light tan color and they were then ready to be split into smaller strips.

Nutrients and Calories of Pine Seeds, Acorns, and Modern Wheat (adapted from Farris 1980)

Plant Material	Protein (%)	Fat (%)	Carbs (%)	Energy Content (calories/3.5 oz)
Sugar pine seed	21.4	53.6	17.5	594
Coulter pine seed	25.4	51	14.4	574
Pinyon pine seed	8.1	23	56.3	450
Black oak acorn	3.8	19.8	64.8	443
Corn flour	7.8	2.6	76.8	361
Wild onion	2.2	0.4	20.8	96
Modern wheat	13.3	2	71	352

Big-cone Spruce (*Pseudotsuga macrocarpa*)

Description: This is a tall tree, 35 to 60 feet tall with drooping branches. The leaves are needle-like, blue-green, and spirally arranged on the branches, but appear to be in a flat spray because the needles are turned at the petiole base. Needles are $\frac{3}{4}$ to $1\frac{1}{2}$ inches long and pointed at the tip. Cones are cylindrical, 4-6 inches long, with 3-fingered bracts overlapping the scales. These bracts are characteristic of the genus.

Ecology & Ethnobotany: This is a common species in the Coulter Pine (*Pinus coulteri*) phase of the California mixed conifer forests, and is typically found on steep north-facing slopes and in ravines (i.e., sites with relatively low fire frequency). At the lowest elevations (~3,500 feet) it occurs as scattered individuals 45-90 feet tall above a closed canopy of Canyon Live Oak (*Quercus chrysolepis*). At higher elevations (~4,500 feet) it becomes much more abundant in a mixed *Pseudotsuga-Quercus* canopy. Although often co-dominant with *Pinus coulteri* and *Quercus chrysolepis*, it is typically found on relatively more mesic (wet) sites with lower fire frequency. Not surprisingly, then, it is also found in riparian habitats as a co-dominant with mesic hardwood species such as Bigleaf Maple (*Acer macrophyllum*) and Cottonwood (*Populus trichocarpa*).

There is no current commercial market for Bigcone Douglas-fir wood due to limited distribution and access. It is heavy, hard, and fine grained but not durable. There is less sapwood than heartwood, and the latter contains pockets of resin. In the past, the wood was used locally for fuel and lumber.

FLOWERING PLANTS: DICOTYLEDONS

Dicotyledonous plants (dicots) can be distinguished by several key characteristics. In dicots, the embryos have two cotyledons (seed leaves), and the leaves are usually net-veined. The flower parts are in fours or fives, rarely in threes, and the plants are herbaceous or woody. Most importantly, the possession of cambium cells distinguishes dicotyledons from the monocotyledons.

MAPLE FAMILY (Aceraceae)

This family consists of two genera and about 200 species distributed worldwide. In general, they are shrubs or trees with opposite leaves that may be simple or compound. The flowers are small, usually appearing before the leaves. The family is of economic importance as a source of timber, ornamentals, and sugar. Maple wood is considered to be heavy, tough, compact, and very hard. Its light brown color with a dense even grain and fine texture makes it one of the best woods for furniture, veneers, and flooring. It is also used in making violins, tool handles, and pianos.

Maple (Acer)

Description: Maples are deciduous trees or shrubs with male and female flowers on the same or separate plants. Flowers are arranged in racemes, corymbs, or umbels. Fruits are winged schizocarps that resemble tiny "helicopters" when blown by the wind. Maples are usually found in moist places in canyons, hills, and along streams from low to high elevations. Of the approximately 15 species of *Acer* native to the United States, three are found in the southern California mountains.

"Quick Key" to the Maples

1. Leaves compound - *A. negundo*
1. Leaves simple – 2

2. Leaves deeply lobed, 4-10 inches across - *A. macrophyllum*
2. Leaves shallowly lobed, $\frac{3}{4}$ - $1\frac{1}{2}$ inches across - *A. glabrum*

Box Elder (*A. negundo*) - Box Elder grows along streams and in moist areas below 6,000 feet. Flowers from March to April.

Bigleaf Maple (*A. macrophyllum*) - Bigleaf Maple is fairly common in canyons and along streams below 5,000 feet. Blooms from April to May.

Mountain Maple (*A. glabrum*) - Mountain Maple is found from 6,500 to 9,000 feet. Blooms from April to May.

Ecology & Ethnobotany: Sap can be harvested in much the same way as the eastern Sugar Maple. To obtain sap, simply bore a small hole into the tree a couple feet above the ground. The sunny side of the tree is usually the ideal spot to bore. Insert a small grooved wooden peg into the hole. This peg  will be the spigot. If the tree is ready to flow, sap will immediately begin to flow after drilling. Hang or place a container under the spigot to collect sap. After collecting sap, seal the hole to protect it from infection and further sap loss while it heals.

Next, the sap must be boiled down because the majority of it is water. Only a small fraction of the original volume collected will be left. You may boil the sap so far as your personal taste dictates. As an alternative to boiling, the collected syrup can be allowed to freeze overnight, which allows the water to separate from the sap. The frozen water can be easily discarded.

The inner bark of all Maples can be eaten in emergencies. A tea made from the inner bark of Box Elder is used to induce vomiting. The young shoots of Mountain Maple can be used as asparagus. The winged seeds of Box Elder can be roasted and eaten.

Native Americans used the young saplings for basketry work. The saplings were split into quarters, as a white wrapping or sewing strand in coiled basketry work. In some places, Maple thickets were intentionally manipulated by burning the old growth to promote new growth. These straight, uniform shoots were highly valued as good basketry material. Maple wood has been used to make snowshoe framing, mush paddles, and other household utensils. Knots and burls on tree trunks can be used for making bowls, dishes, and other items. Gum from the buds was mixed with animal fat and used as a hair tonic. Inner bark can be shredded and twisted into a coarse rope.

AMARANTH FAMILY (Amaranthaceae)

The Amaranth Family contains more than 60 genera and 900 species distributed worldwide. Many are weedy species of little economic importance. Some species of *Amaranthus* are cultivated for their red pigmentation. The family name originates from the Greek *amarantos*, which means unfading, possibly alluding to the "everlasting" quality of the papery perianth parts.

Pigweed, Amaranth (*Amaranthus*)

Description: Approximately 16 widespread species of Amaranthus occur in southern California. In general, they are herbaceous annuals with small greenish flowers, and alternate entire or wavy margined leaves. Pigweeds occur in many different habitats and often hybridize, making identification difficult.

Ecology & Ethnobotany: Used by Native Americans, the dried small black seeds of Amaranth have been found in many archeological remains. Seeds of all species can be eaten whole as a cereal or ground into meal, and made into cakes. The seeds are best collected in summer when the plants are fully mature. To free the seeds from their husks, rub the seed clusters between your hands. You can then winnow the seeds if there is a breeze, or if air is calm, slowly pour the seeds out of your hands and blow the chaff away. The seeds contain approximately 15 grams of protein per 100 grams, more than is found in rice and corn, and equal to, if not surpassing that found in wheat. When ground into a flour, Amaranth has a distinctive flavor that is a bit strong used alone. We  find it is better when mixed with other flours for breads and pancakes.

The highly nutritious Amaranth contains more fiber and calcium than any other cereal grain in addition to a wide spectrum of vitamins and minerals, including Vitamins A and C, calcium, magnesium, and iron. Amaranths is rich in the amino acid lysine, a product scarce in true cereal grains, thereby providing a more balanced source of protein.

The edible young shoots and leaves have a pleasant taste if eaten as a potherb soon after collection. Since the plants can accumulate nitrates, it is wise not to consume large quantities where nitrate fertilizers are used. Livestock losses have occurred as a result of excessive Amaranth consumption. The leaves of Amaranth contain oxalic acid which binds with calcium restricting its absorption by the body. As long as your diet contains plenty of calcium from other sources, eating Amaranth and other vegetables that contain oxalic acid (e.g., Spinach, Wood Sorrel), should not cause any health problems.

Amaranth also has astringent properties and can be used for treating diarrhea, excessive menstrual flow, hemorrhaging, and hoarseness. Amaranth is also helpful in treating mouth and throat inflammations and sores, and in quelling dysentery and diarrhea. Steeping dried leaves in boiling water (the more leaves steeped, the stronger the tea) was considered a valuable remedy.

In the Mid-west, several species of Amaranth are being grown as agricultural crops for use in cereals and bread. It is photosynthetically efficient and produces a high yield of both greens and seeds. Amaranth was an important food in the past and may become an important one for the future.

SUMAC or CASHEW FAMILY (Anacardiaceae)

There are approximately 79 genera and 600 species in this family. Products originating from the Sumac Family include resins, oils, lacquers, edible fruits, ornamentals, and tannic acid. Although many members of this family produce edible fruits, the resinous oils can produce extreme dermatitis in sensitive individuals. The family contains the infamous Poison Ivy (*Toxicodendron rydbergii*), Poison Oak (*T. diversilobum*), and Poison Sumac (*T. vernix*).

Skunkbrush (*Rhus trilobata*)

Description: Approximately 60 species of this genus occur worldwide. Skunkbrush is a shrub that occurs on dry sunny slopes, has compound leaves with three leaflets, of which the middle leaflet is largest. Skunkbrush has small yellow-green flowers that bloom before the leaves come out. The red-orange fruits are sour tasting. The genus name comes from the Greek *rhous*, which is the name of a bushy sumac.

Ecology & Ethnobotany: The berries of Skunkbush can be eaten raw, or soaked in cold water to make a refreshing drink. Malic acid (the cause of tartness in apples) flavors the sour fruits of the sumacs.

Tannic acid which is present in all parts of the plants, can be used in tanning leather (hides). The leaves, branches, and fruits provide colorfast dyes for wool. The stem produces a yellow dye and the berries a tan or beige dye. Since tannic acid acts as a natural mordant (dye fixer), the fiber does not need to be treated with other chemicals.

The slender, flexible branches of Skunkbush can be used for weaving baskets as they are somewhat vine-like. The branches can also be used as chew-sticks to clean teeth and massage gums. Take a small stem several inches long, remove the outer bark and chew on the tip to soften fibers. Since some people may have an allergic reaction to the oils of sumac, it is recommended that this be done sparingly.

Sumac-ade

The drink, called "sumac-ade," can be made from the juice of the fruit. To make this drink, simply collect several clusters of the ripened berries, clean off any excess herbage, then bruise the fruits slightly, and extract the juice by soaking them in enough cool to warm water to cover the fruit fully. Since sumacs contain high levels of tannic acid, use cool water rather than hot water so little or no tannic acid will be extracted. Soak the berries for about 20-30 minutes, strain through a cloth, and drink.

Poison Oak (*Toxicodendron diversilobium*)

Description: Poison Oak is a low shrub or woody vine found in waste places, hillsides and rocky ravines in the lower elevations of the State. The leaves are compound with three green, oval shaped, and pointed leaflets, which turn bright red or orange in the fall. The white flowers arise from the leaf axils, and the fruits are white berries.

Ecology & Ethnobotany: The foliage of Poison Oak is poisonous, causing contact dermatitis. The actual skin irritant is found in the sap. The itchy or painful rash that develops from contact with the sap is greatest in spring and summer when the sap is abundant and the plant is easily bruised. Shortly after contact, the symptoms include itching, burning, and redness. Small blisters may appear after a few to several hours. Severe dermatitis, with large blisters and local swelling, can remain for several days and may require hospitalization. Secondary infections may occur when the blisters are broken. To help alleviate itching immediately, thorough washing with soap and water after contact is recommended.

Since droplets of the irritating chemical can be carried in smoke on dust particles and ash, do not burn Poison Oak. Smoke carries the oil and can produce a rash over the whole body. If inhaled, a rash can develop in the throat, bronchial tubes, and lungs. The oil can even spread through the body in the blood stream. Although Poison Oak can cause havoc for humans, robins, cedar waxwings, flickers, woodpeckers, and other birds relish the berries.

CARROT FAMILY (Apiaceae)

Approximately 300 genera and 3,000 species are found in this family. About one-quarter of the genera and 10% of the species are native to the United States. The Carrot Family is of considerable economic importance because of numerous food plants, condiments, ornamentals, and poisonous species. Some familiar members of this family include Carrot (*Daucus*), Parsnip (*Pastinaca*), Celery (*Apium*), Anise (*Pimpinella*), and Parsley (*Petroselinum*).

Please Note: No wild members of this family should be eaten until they have been accurately identified. Correct identification usually requires the mature fruit called a schizocarp (a dry fruit that splits into two halves).

Angelica (*Angelica tomentosa*)

Description: This is a tall, often aromatic, taprooted perennial with leaves that are mostly twice divided into broad, toothed or lobed leaflets. The flowers are mostly white. The fruit is narrowly elliptical to nearly orbicular with winged ribs on the outer face. The species is found in rather moist and shaded areas below 6,000 feet, from the mountains in San Diego northward. Blooms June to August.

Ecology & Ethnobotany: While there are a number of edible species of Angelica, we have not found any information regarding the edibility of this species. Therefore, the internal use of any Angelica species in southern California is not recommended until studies have been conducted concerning their toxicity, and because they superficially resemble the poisonous Water Hemlock (*Cicuta maculata*). All species of Angelica contain furanocoumarins, which increase skin photosensitivity and may cause dermatitis.

Very little is known about the medicinal aspects of this species. Angelica has been described, in general, as an antispasmodic, expectorant, diaphoretic, diuretic, an effective astringent to the stomach lining, and a menstrual stimulant that helps reduce cramps. Angier (1978) describes a volatile oil in eastern species of Angelica, that was used to treat colic and digestive gas.

Poultices from the mashed roots of Angelica were applied for arthritis, chest discomfort, and pneumonia.

Warning: Angelica closely resembles the highly poisonous Water Hemlock (*Cicuta douglasii*).   Positive identification of the plants is paramount.   The identification of young plants of Angelica and Water Hemlock can be made by examining the leaf venation. The leaf edges of Angelica are serrate and pinnately divided into opposing pairs, like Water Hemlock, but the leaf veins extend from the midribs to the outer tips of the serrations. Water Hemlock has leaf veins terminating within the notches of the serrations.

Water Hemlock (*Cicuta maculata*)

Description: This species is found in marshes and along the edges of streams and ponds from low to mid-elevations (below 8,000 feet). Water Hemlock is a stout perennial from fleshy, fascicled roots. Leaves are 1-2 times pinnately divided into narrowly lance-shaped, sharply toothed leaflets.   The veins of the leaflets terminate at the notches between the teeth. Numerous white to greenish flowers are arranged in compound umbels. Fruits are slightly flattened with thickened ribs on the faces. The bruised foliage produces a musky odor. Flowers from June to September.

Ecology & Ethnobotany: These are extremely poisonous plants if ingested. In fact, Water Hemlock has been described as the most violently poisonous vascular plant in North America. The whole plant contains cicutoxin, a resin-like substance that depresses the respiratory system, with the root being particularly dangerous. A single mouthful of the plant is capable of killing an adult. Water Hemlocks have been used throughout the ages to execute criminals and kings. Many children have been fatally poisoned by blowing into whistles made from hollow stems of Water Hemlock. In Oregon, Native Americans soaked arrows in Cicuta juice, rattlesnake venom, and decayed deer liver to poison arrow tips for hunting.

Strike (1994) indicates that Water Hemlock roots were mashed and smeared on a hot stone to relieve pain in sore arms or legs. The mashed root was then pressed against the sore arm or leg.

Following is a graphic description of poisoning due to the ingestion of Water Hemlock in Europe. If anything, it should instill into the minds of wild food gatherers the need to positively identify a plant species before eating it.

"When about the end of March 1670, the cattle were being led from the village to water at the spring, in treading the river banks they exposed the roots of this Cicuta (water hemlock) whose stems and leaf buds were now coming forth. At that time two boys and six girls, a little before noon, ran out to the spring and the meadow through which the river flows, and seeing a root and thinking that was a golden parsnip, not through the bidding of any evil appetite, but at the behest of wayward frolicsomeness, ate greedily of it, and certain of the girls among them commended the root to others for its sweetness and pleasantness, wherefore the boys, especially, ate quite abundantly of it and joyfully hastened home; and one of the girls tearfully complained to her mother she had

been supplied meagerly by her comrades, with the root.

Jacob Maeder, a boy of six years, possessed of white locks, and delicate though active, returned home happy and smiling, as if things had gone well. A little while afterwards he complained of pain in his abdomen, and, scarcely uttering a word, fell prostrate to the ground, and urinated with great violence to the height of a man. Presently he was a terrible sight to see, being seized with convulsions, with the loss of all his senses. His mouth was shut most tightly so that it could not be opened by any means. He grated his teeth; he twisted his eyes about strangely and blood flowed from his ears. In the region of his abdomen a certain swollen body of the size of a man's fist struck the hand of the afflicted father with the greatest force, particularly in the neighborhood of the ensiform cartilage. He frequently hiccupped; at times he seemed to be about to vomit, but he could force nothing from his mouth, which was most tightly closed. He tossed his limbs about marvelously and twisted them; frequently his head was drawn backward and his whole back was curved in the form of a bow, so that a small child could have crept beneath him in the space between his back and the bed without touching him. When the convulsions ceased momentarily, he implored the assistance of his mother. Presently, when they returned with equal violence, he could not be aroused by no pinching, by no talking, or by no other means, until his strength failed and he grew pale; and when a hand was placed on his breast he breathed his last. These symptoms continued scarcely beyond a half hour. After his death, his abdomen and face swelled *without lividness except that a little was noticeable about the eyes. From the mouth of the corpse even to the hour of his burial green froth flowed very abundantly, and although it was wiped away frequently by his grieving father, nevertheless new froth soon took its place (Jacobson 1915)."*

Poison Hemlock (*Conium maculatum*)

Description: Poison Hemlock is a biennial with a stout taproot and a disagreeable odor when crushed. The stems are purple-blotched and hollow, and the leaves are pinnately dissected with a lacy appearance to them. The flowers are white in compound umbels, and the fruits are egg-shaped, flattened with prominent, wavy ribs. The plant is usually found in disturbed sites and waste places at low elevations (below 5,000 feet). Blooms April to September.

Ecology & Ethnobotany: This is an extremely poisonous plant. Death is said to result from the ingestion of the leaves, roots or seeds. The most famous use of Poison Hemlock was by the ancient Greeks as a humane method of capital punishment. It is said to be quite painless and the recipient's mind remains clear to the end. Introduced from Europe, Poison Hemlock has established itself as a common weed. Socrates was said to be killed by the plant in 399 BC when he was forced to drink it.

Spring Parsley (*Cymopteris multinervatus*)

Description: The leaves are 2-4 times pinnately divided into small ultimate segments, and the flowers are yellow or white in terminal, compound umbels. The round fruits have winged ribs on the outer faces. This species occurs on dry slopes from 3,500 to 6,000 feet. Blooms from March to April.

Ecology & Ethnobotany: All species produce edible roots. We found the older roots more fibrous than the younger ones. The root can be used in stews or it can be boiled or roasted in a pit, mashed and dried into cakes. When dried it will keep indefinitely. During the Lewis and Clark expedition this was known as kouse (bread of cows). The old roots can also be used as an effective insect repellant when boiled. Just sprinkle the tea around camp and in sleeping areas.

The upper parts of the plants have been used raw or as potherbs. If cooked, they will require several changes of water. The seeds of some species are edible when ground and used as flour.

Rattlesnake Weed (*Daucus pusillus*)

Description: This is a pubescent annual with pinnately compound leaves. The flowers are white and the fruit is oblong to ovoid and flattened dorsally. These plants resemble Poison Hemlock (*Conium maculatum*), but are readily distinguished from each other in that Daucus has stems and leaves that are distinctly hairy. This is a common species on dry slopes below 5,000 feet, especially after fire or disturbance. Flowers April to June.

Ecology & Ethnobotany: The crushed seeds have been used as a contraceptive and herbal "morning after pill" for at least 2,500 years. Rattlesnake Weed has been shown to be successful for use as a contraceptive in laboratory trials, and is used in some areas today.

The Costanoan Indians used *D. pusillus* to reduce fevers, heal snake bites, cure colds, and as a general blood medicine. As a poultice, the plant was used on bruises and swellings (Strike 1994).

Another species that may occur in the southern California mountains is Queen Anne's Lace (*D. carota*). It is common at the lower elevations along roadsides, in fields, pastures, waste places, and moist clearings. Introduced from Europe, it is the ancestor of the cultivated carrot. The first year's roots can be prepared like garden carrots. We found the older roots tough and stringy. The roots can also be dried and roasted and then ground for use as a coffee substitute. The plant was used extensively by many Native Americans and should be kept in mind as an emergency food.

A tea made from the root has been traditionally used as a diuretic to prevent and eliminate urinary stones and worms. Laboratory studies confirm the bactericidal, diuretic, and worm expelling properties. The seeds of this species are also considered to be a folk "morning after" contraceptive.

"Nature does nothing in vain."

-Sir Thomas Browne
(said to be "the only undisputed
axiom in philosophy")

Cowparsnip (*Heracleum lanatum*)

Description: Cowparsnip is a stout perennial up to or more than seven feet tall. The lower leaves are three lobed, resembling a maple leaf up to 14 inches long. The white flowers are in compound umbels and the fruits are egg-shaped with only the marginal ribs winged. It is usually found in moist soils around streams, seeps, and avalanche chutes up to subalpine environments. The genus is named for Hercules, who is reputed to have used it as a medicine. Blooms April to June.

Ecology &

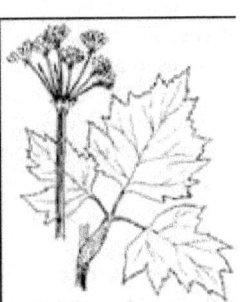

Ethnobotany: The young stems of Cowparsnip can be peeled and eaten raw, but are best when cooked. The hollow base of the plant can be cut into short lengths and used as a substitute for salt by eating or cooking with other foods. The young leaves are also edible after cooking, but we find them not very tasty. The leaves can also be dried and burned, and the ashes used as a salt substitute. Strong and bitter tasting, the cooked roots are said to be good for digestion, as well as for relieving gas, and cramps. In our experience, some plants are much more palatable than others.

The seeds can be sparingly added to salads for seasoning. However, you should be aware that the mature, green seeds have a mild anesthetic action on tissues in the mouth. When gently chewed and sucked, they will numb the tongue and gums in a manner similar to clove oil.

Medicinally, root pieces were placed in tooth cavities to stop pain. An infusion for a sore throat can be made by soaking the mashed root in water. Root tea was used for colic, cramps, headaches, sore throats, and flu.

The leaves of Cowparsnip are large enough to be used as a toilet paper substitute and as a mild insect repellent. However, since furanocoumarin is present in the sap and the outer hairs, it may be a problem for people with sensitive skin. When the sap comes in contact with the skin in sensitive people it causes a type of "sunburn" effect (i.e., redness, blistering, and running sores) when exposed to light. As a poultice, the leaves were used externally for sores, bruises, and swellings.

The older stems, before the flower cluster unfolds, can be peeled and the inner tissue eaten raw or cooked. While it is edible, it does have an unpleasant taste. Cooking it in a couple changes of water usually helps the taste and digestibility. In any case, Cowparsnip is considered to be an excellent survival plant in the mountains. Strike (1994) indicates that Cowparsnip has been probably the most intensively used springtime green among Native Americans in Canada.

Warning: Do not confuse this plant with other species in the same family that are highly toxic (i.e., *Cicuta* and *Conium maculatum*)

Biscuitroot (*Lomatium*)

Description: These are perennial plants with thick roots and leaves that are divided several times from the base. The white, yellow, pink, or purplish flowers are in compound umbels. The fruits are flattened and elliptical to oval in shape, and the margins may or may not be winged. Most species are found in dry ground or rocky situations. The genus name means "small border", alluding to the wings of the fruit. The following two species are known to occur in the southern California mountains.

"Quick Key" to the Biscuitroots

1. Bractlets broad, rounded to obovate; flowers yellow - *L. nevadensis*
1. Bractlets narrow, linear to lanceolate; flowers purple or yellow - *L. dissectum*

    Fern-leaf Lomatium (*L. dissectum*) This perennial herb occurs in rocky soils from the chaparral and woodlands to subalpine (500 to 9,000 feet).
    Lomatium (*L. nevadensis*) This Lomatium grows on dry slopes between 5,000 and 9,000 feet in the San Bernardino and San Gabriel Mountains. Flowers from April to July.

    Ecology & Ethnobotany: All Biscuitroot species have edible roots and were an important staple among many Native Americans. They can be eaten raw or cooked, or dried and ground into flour. The flour can then be kneaded into dough, flattened into cakes, and dried in the sun or baked. Some of the species we have tried were too resinous to enjoy. Personal taste will guide one to choose the more palatable species.
    The green stems can be eaten after boiling in the springtime, but as summer progresses, they become tough and fibrous. A tea can be brewed from leaves,

stems, and flowers. The tiny seeds nutritious raw or roasted, can be dried and ground into meal.

The plants are rich in Vitamin C. Seeds were chewed for colds and sore throats, and sap from the roots was used to treat cuts and sores. A poultice of pulverized roots was applied to a newborn baby's umbilical cord to facilitate healing. Roots were also chewed to relieve sore throats. Recent studies have shown that Fernleaf Biscuitroot has an ability to kill certain forms of influenza virus, especially those that infect the respiratory tract. It also has other anti-microbial and immuno-stimulating qualities.

Caution: As with any member of the Carrot Family, positive identification is important before consumption. Strike (1994) indicates that some indigenous peoples considered the mature stalks, leaves, roots, and flowers of *L. dissectum* as poisonous. In fact, the roots were used as a fish poison and insecticide by some Native People in the West. The plant contains phototoxic compounds of the furanocoumarin group and one or more of these compounds is responsible for the fish poison and insecticidal properties found in the chocolate tipped roots.

Sweet Cicely (*Osmorhiza*)

Description: These two species are herbaceous perennials from stout roots, with leaves twice divided into 3's. The flowers are borne in open, compound umbels that arise from leaf axils. The fruit is spindle-shaped and compressed along the sides. In general, they occur on moist slopes, open areas, and forests.

"Quick Key" to the Sweet Cicelies

1. Rays subtended by bracts; flowers greenish-yellow -
O. brachypoda
1. Rays not subtended by bracts; flowers greenish-white
- *O. chilensis*

> **California Sweet Cicely (*O. brachypoda*)** This species grows in shaded woods below 8,500 feet, and flowers from March to May. Flowers are greenish-yellow in color.
> **Sweet Cicely (*O. chilensis*)** Sweet Cicely grows in woods below 8,000 feet throughout southern California. Flowers from April to July. Flowers are greenish-white in color.

Ecology & Ethnobotany: The leaves of Sweet Cicely, also known as "dryland" parsnip, can be boiled and then eaten. The roots were dug in the spring and either pit cooked or boiled as a vegetable. To us the taste is reminiscent of baby carrots.

The roots of a related species, *Osmorhiza occidentalis* (Western Sweetroot), taste and smell like licorice or anise, and can overwhelm the taste buds if eaten in large amounts. In small quantities, Western Sweetroot can liven up the taste of teas (or meals) that are otherwise bland or unpleasant.

Uses of other related species include a poultice from

roots for boils, cuts, sores, and wounds, and root tea for sore throat and upset stomach.

Caution: Western Sweetroot resembles the very poisonous Water Hemlock (*Cicuta douglasii*), but the strong smell of anise gives it away as Sweetroot. Also, the venation of Water Hemlock is unique among the Apiaceae (Carrot Family). Sweetcicely can be confused with Baneberry (*Actea rubra*). However, Baneberry is easy to distinguish from Sweetcicely by the cluster of red or white berries.

Wild Parsnip (*Pastinaca sativa*)

Description: Wild Parsnip is a biennial with stout, leafy stems that arises from a large taproot. The basal leaves are once pinnately compound into usually 9-13 lance- to egg-shaped leaflets. Flowers are yellow and in compound umbels. The fruits have fine ribs on the outer face. It can be found in damp disturbed areas at the lower elevations. It apparently is localized as a naturalized plant in San Jacinto Mountains and San Gabriel Mountains. Blooms from June to July.

Ecology & Ethnobotany: This introduced plant from Europe is the wild form of the cultivated parsnip and can be prepared and eaten in the same way. The taproot from first year plants can be eaten raw or boiled until tender. The root of the plant was used by many people, from the ancient Romans to Native Americans. In Holland, the plant was used in soups. The Irish made a beer by boiling the roots with water and hops, then allowed the mixture to ferment. A tea from the roots was used by Native Americans to treat sharp pains. The roots were also used as a poultice on inflammations and sores.

Caution: Due to the presence of xanthotoxin, the plant may cause photodermatitis. The symptoms are much like those from exposure to Poison Ivy, but of longer duration. Xanthotoxin is used to treat psoriasis and virtiligo. Therefore, avoid contact and exposure to sunlight.

Yampah *(Perideridia)*

Description: These are biennial or perennial herbs with fascicled tuberous roots and pinnate leaves. The calyx-teeth are well-developed. The petals are white or pinkish, the stylopodium conic. The fruit is nearly terete or somewhat flattened laterally.

"Quick Key" to the Yampahs

1. Fruit globose in shape - *P. gairdneri*
1. Fruit oblong in shape - *P. parishii*

Common Yampah (*P. gairdneri*) This species occurs in wet places below 11,000 feet. Flowers from June to July.

Perideridia (*P. parishii*) Perideridia grows in moist meadows from 3,500 to 11,000 feet. Flowers from July to September.

Ecology & Ethnobotany: All of the species within this genus are edible. They were an important food of many indigenous peoples from British Columbia to California and the Great Basin region. The raw finger-like roots have a pleasant, nutty flavor when eaten raw, and resemble carrots when cooked. They are best when dug up before the flowers appear. The roots should be washed and peeled before cooking. They can be easily dried and will keep well for future use. When dried, the roots can be pounded and ground into flour or mashed into cakes. The seeds may be parched and ground or eaten whole.

Snakeroot (*Sanicula*)

Description: They are erect perennials with 3-5 lobed leaves. The flowers are in compound umbels and the fruits are flattened laterally, densely covered by bristles. Two species occur in the southern California mountains.

"Quick Key" to the Snakeroots

1. Flowers purple - *S. bipinnatifida*
1. Flowers yellow - *S. graveolens*

Purple Sanicle (*S. bipinnatifidia*) Purple Sanicle is fairly common on open slopes from 3,500 to 6,000 feet. The species flowers from March to May.

Sanicle (*S. graveolens*) This species grows in open forests from 4,000 to 8,000 feet throughout southern California. Sanicle flowers from April to June.

Ecology & Ethnobotany: The herbage of both species is known to contain various alkaloids and should therefore be regarded as inedible.

Ranger's Button (*Sphenosciadium capitellatum*)

Description: This is a stout, perennial plant that grows 20 to 64 inches tall. The leaves are large, 1 to 2 times pinnate on swollen petioles. The leaflets are linear, oblong, 3/8 to 3/4 inch long. The white flowers occur in compact umbels that are ball-like in appearance. The fruits are flattened dorsally, about $\frac{1}{4}$-inch long, woolly and with lateral ribs winged. Ranger's Button can be found in moist or wet meadows, bogs or streambanks between 3,000 to 10,000 feet. Flowers in July and August.

Ecology & Ethnobotany: The roots were chewed by the Maidu Indians in California to relieve sore throats, and a root decoction was used to treat bronchial problems. An infusion of the roots was used by the Maidu and Paiute Indians to repel lice.

Tauschia (*Tauschia*)

Description: These perennial herbs from taproots or tubers have leaves that are pinnately compound. The yellow, white, or purplish flowers occur in loose compound umbels. The genus name honors I.F. Tausch, a 19th century botanist. Two species of Tauschia can be found in the southern California mountains.

Tauschia (*T. arguta*) This species grows on dry, sandy, or gravelly areas below 7,000 feet. Flowers

---

from April to June. The leaves of this species are once pinnate.

Tauschia (*T. parishii*) This species grows on dry slopes from 4,000 to 9,000 feet. Flowers from May to July. The leaves of this species are two times pinnate.

Ecology & Ethnobotany: The boiled root of *T. parishii* was used to relieve internal pains by the Kawaiisu. For toothaches, the fresh root was mashed or dried roots were pulverized and smeared on hot rocks. By placing the cheek of the toothache sufferer directly on the mashed root, the toothache was cured.

DOGBANE FAMILY (Apocynaceae)

This is a large family of about 200 genera and 2,000 species that are mostly found in the tropics. Nearly all of the members within this family are poisonous and usually have milky juice. Some of the well-known genera are ornamentals such as *Vinca minor* (Periwinkle) and *Nerium oleander* (Oleander). In recent years *Rauwolfia serpentina* (Indian Snakeroot), a tropical tree, was found to yield a wonder drug used in the treatment of high blood pressure.

Amsonia, Woolly Bluestar (*Amsonia tomentosa*)

Description: This is a perennial herb that grows from 2,000 to 6,000 feet on dry slopes. Flowers from March to May. The genus name honors Dr. Charles Amson, an 18th century resident of Virginia.

Ecology & Ethnobotany: Woolly Bluestar was used by the Zuni Indians made a poultice of the root and applied it with mush ceremony to a rattlesnake bite. The Paiutes made carrying straps from the bark of Amsonia. The straps were braided three-ply twisted.

Spreading Dogbane (*Apocynum androsaemifoilium*)

Description: This is a glabrous, erect perennial up to 18 inches tall. The leaves are opposite, ovate, drooping, dark green glabrous above, and lighter below. Flowers occur in terminal clusters, and the calyx is 5-cleft. The white with pinkish veined corolla is bell-shaped and 5-lobed. The fruit is a pod, about $\frac{1}{2}$ inch long. Spreading Dogbane grows on dry flats and slopes from 5,000 to 9,500 feet. Dogbane flowers from June to August. *Apocynum* is Greek meaning "noxious to dogs".

Ecology & Ethnobotany: The primary use of Dogbanes is for fiber. The stem fibers are strong and can be used for rope making, mats, baskets, bowstrings, fishing lines and nets, sewing, animal trap triggers, snares, cordage for bow and drill fire making, and general weaving. One of the easiest ways to isolate the fibers is to soak the stems in water. Archeologists in Utah have discovered nets made with the fiber dating back to about 5000 B.C. Many Native American tribes used Dogbane to make rabbit-catching nets. Some of these nets were about 200 feet long, three to four feet high, with a 3-inch opening. The nets were propped on sticks across level ground. The men formed a line some distance away and advanced toward the nets, beating the brush with sticks, and driving the rabbits into the net.

Dogbanes should be considered poisonous to humans if ingested. However, some authorities have indicated that the small seeds can be parched, ground into a meal to make fried cakes. Strike (1994) believes the seeds eaten were that of *A. pumilum* (Mountain Dogbane). *Apocynum pumilum* is a subspecies of *A. androsaemifolium*.

Dogbanes were extensively used as medicine by aboriginal peoples. They contain highly toxic glycosides and resins with cymarin and apocannocide being major medicinal constituents found throughout the plants. Research indicates that the cardiac glycosides may be useful in treating malignant tumors. Millspaugh (1974) describes Spreading Dogbane as an emetic without causing nausea, a cathartic, and quite powerful diuretic and sudorific; it is also an expectorant and antisyphilitic.

"Fiber plants are second only to food plants in terms of their usefulness to humans and their influence on the advancement of civilization. Tropical people use plant fibers for housing, clothing, hammocks, nets, baskets, fishing lines, and bow strings. Even in our industrial society, we use a variety of natural plant fibers ... In fact, the so-called synthetic fibers now providing much of our clothing are only reconstituted cellulose of plant origin."

—Mark Plotkin (1988)
"Biodiversity"

GINSENG FAMILY (Araliaceae)

There are about 50 genera and 500 species worldwide. Though they are of limited economic importance, several species are used as ornamentals. One species, American Ginseng (*Panax quinquefolius*) is the famous medicinal panacea and "mind enhancer".

Elk Clover, Wild Sarsaparilla (*Aralia californica*)

Description: This is a large perennial with stems that grows up to 9 feet tall. The roots have a milky juice, and the leaves are 2-times pinnate. The leaflets are toothed and ovate. The small flowers occur in a many-flowered umbel with petals less than 1/8-inch long. he berry-like fruit is red when young, becoming black with age. This species grows on moist and shaded slopes below

5,500 feet from Orange County northward. Flowers from June to August.

Ecology & Ethnobotany: A poultice from the rhizomes of a related species, *A. nudicaulis*, can be used to treat burns and sores. As a tonic, it was a diuretic that lowers fever. As a tea, Wild Sarsaparilla is rather pleasant tasting. Internally, it was used for coughs and purifying blood. The long rootstalk is often used as an ingredient of rootbeer. Fernald and Kinsey (1958) indicate that Native Americans relied for long periods of time on these roots during wars or when hunting. It has been used as a substitute for true Sarsaparilla (*Smilax officinalis*). The roots and rhizomes of another related species, *A. racemosa*, have been used to treat rheumatism, coughs, and backaches.

<u>MILKWEED FAMILY (Asclepiadaceae)</u>

About 250 genera and 2,000 species are found in this family worldwide. Milky sap in the stems, leaves, and flowers inspired the common name for the Milkweed Family. The family is of moderate economic importance as a source of ornamentals, latex, fibers, poisonous plants, and a few food plants. The sap contains latex, and in a few species, it may yield industrially important hydrocarbons. The flowers are 5-parted. Pollination involves an insect literally pulling up the pollen mass from the anthers and directly depositing it on the stigma.

Milkweed (*Asclepias*)

Description: These are erect or decumbent herbs from deep perennial roots. The leaves are opposite or whorled and the corolla is deeply 5-parted with the segments reflexed. The corona hoods each have an incurved horn within. The name *Asclepias* refers to Asklepios, the Greek god of medicine.

California Milkweed (*A. californica*) California Milkweed is common on dry slopes below 7,500 feet in the chaparral and the yellow pine forest habitats. Flowers from April to July.

Milkweed (*A. eriocarpa*) Milkweed is common on dry washes and slopes below 7,000 feet. Flowers from June to August.

Narrow-leaf Milkweed (*A. fascicularis*) This species occurs in dry places mostly below 7,000 feet in many plant communities, usually on the desert side of the mountains.

Showy Milkweed (*A. speciosa*) This species occurs in many habitats, including roadsides and disturbed areas below 6,000 feet.

Woolly Milkweed (*A. vestita*) This species is found on dry slopes below 5,000 feet in the Transverse Ranges northward. Flowers April to June.

Ecology & Ethnobotany: Almost every book on edible plants in the United States lists Milkweeds as being edible. It should be noted that in most cases they are referring to the eastern species of *Asclepias syriaca* which does not occur in California, and the western species of *A. speciosa* (Showy Milkweed) and *A. asperula* (Spider Milkweed). These latter two species also contain the cardiac glycosides that can cause severe poisoning if not properly prepared or cooked.

With respect to the two western species mentioned above, Harrington (1967) suggests gathering plants when they are six inches tall and then boiling for 15-20 minutes in at least 2-3 changes of water. We have tried 5 to 7 changes of water and the plants were still bitter!! The unopened flower buds can be served like broccoli by boiling in at least three changes of water.

A strong fiber can be obtained from the inner bark to make rope, fishing line, clothing, and nets. Archeologists have discovered clothing that was made from the fibers more than 10,000 years ago. The silky floss found in mature Milkweed seed pods were used in

 making candlewicks, and the fiber can be spun like cotton. The floss is buoyant and water resistant and makes a good insulator. During World War II, schoolchildren in Canada harvested Milkweed floss from the wild for the United States Navy's use as a substitute for kapok in life vests. The dried pods were used as utensils. The sap was used as an adhesive.

Milkweeds contain asclepain in their plant parts and sap. Asclepain is a proteolytic enzyme that gives credence to the old pioneer remedy of applying the white juice daily to get rid of warts. However, some Native American tribes used to collect the milk of A. *speciosa* and roll it in hand until it became firm enough to chew as gum, but it was not swallowed.

Milkweeds have been used in folk medicine for hundreds of years. The powdered root of several species is reported to have been used to treat wounds, pulmonary diseases, rheumatism, and gastrointestinal problems, among other ailments. Many modern medicines were originally derived from poisonous plants. Perhaps research will validate some of the medicinal uses of Milkweeds and provide us with new medicines from the old (Lewis and Elvin-Lewis 1977).

Additionally, Milkweeds have the potential to furnish an exciting array of products for industry and home. In the future, as petroleum products dwindle, perhaps we will find ourselves taking a closer look at the possibilities of cultivating Milkweeds for fiber, hydrocarbons, and medicines.

Warning: Milkweeds can be confused with other plants producing milky juice such as Dogbane (*Apocynum*), which is also considered to be poisonous. Additionally, some species of Milkweed at certain stages are poisonous to animals and could affect humans when eaten raw.

Wilderness Cordage

The survival and continued existence of primitive humans was as much dependent on fiber as on food. The cordage made from the fibers of wild plants can used to make blankets, sandals, baskets, clothing, nets for fishing, and snares for capturing small game animals. In a wilderness situation, you will be surprised how important a piece of string or cordage can be. There are many species of plants in southern California that have fiber in the stem, leaves, or bark that can in one way or another be used as cordage. Some of the species discussed in this handbook include milkweed (Asclepias), dogbane (Apocynum), sagebrush (Artemisia), cottonwood (Popular), willow (Salix), juniper (Juniperus), thistle (Cirsium), sunflower (Helianthus), yucca (Yucca), and nettle (Urtica). There are probably other species that can be used, and finding out which ones will be a matter of experimentation.

One of our favorite activities when teaching wilderness survival courses is to have participants twist about 50 feet of cordage. After accomplishing that task, students appreciate how much easier it is to go to the hardware store and buy it.

Twisting cordage is relatively easy once the fiber has been extracted from the plant. In most cases, the fiber is located in the outer part of the plant stem. The fibers can be removed by rubbing the stem between the hands or by carefully pounding it with a rounded rock or mallet. It is important to not to break the length of the fiber while doing this. This should result in soft, thread-like fibers.

To twist the fiber into a short piece of cordage, simply roll the length of fiber down your leg with an open palm until it is rounded and reasonably uniform in diameter. However, if longer cordage is required to be used as fishing line, sewing, nets, or clothing, it will be necessary for you to twist and splice. The following directions are for right-handed people. If you are left handed, simply reverse the process.

With the strand of fiber in your left hand, bend it in half. There should be two uneven lengths *hanging down. Pinch the loop you've created with the left thumb and forefinger. With your right hand, grab the strands on the outside, twist it in the outward direction about half-way, and then lay it over the inside strand. Move your left thumb and forefinger down to hold it together. Again, with your right hand, grab the new outside strand, twist it out, and again lay it over the inside strand and reposition your left thumb and forefinger down to hold it together. Repeat this a few more times. When you are about two to three inches from the end of the shorter strand, take another length of fiber and lay it on the shorter piece and twist it as though it were part of the shorter strand. Continue twisting as before until you reach the 2-inch mark with the other strand. Again, attach a new strand and twist it as part of the new one. This is called splicing. If you are doing this for the first time, you'll soon realize that you have muscles in your fingers you never knew you had.*

SUNFLOWER FAMILY (Asteraceae)

This is a very diverse family with over 20,000 species, and it is the second largest plant family in the world. The Sunflower Family contains many economically important crop plants such as sunflowers, lettuce, and artichokes. Numerous edible and useful composites (members of the Sunflower Family) are found in southern California.

While the family is considered by many botanists as a "difficult" group, composites, in general are relatively easy to recognize. The small flowers are arranged in heads that at first appear to be an individual

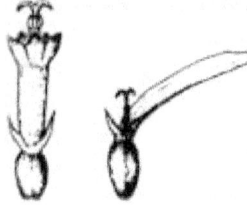 flower, although it may actually consist of several to hundreds of florets (little flowers). Each flower has an inferior ovary, 5 stamens fused at the anthers, and 5 fused petals. The flowers at the center of the head are disk flowers, while the peripheral ones are called ray. Surrounding the head are a series of bracts called the involucre. The calyx, if present (called the pappus and usually modified into thin hairs for dispersal) crowns the summit of the ovary in the form of awns, capillary bristles, scales or teeth. Nearly all composites are herbs or shrubs. The pollen of many composites is allergenic. The colorful flowers of many species produce yellow and orange dyes.

Yarrow (*Achillea millefolium*)

Description: This is a strongly scented perennial herb with alternate leaves that are finely dissected and appear feathery. The white or sometimes yellow flowers are borne in a flat-topped corymb. Yarrow is widespread and can be found in a variety of habitats from low elevations to above timberline. The generic name honors Achilles. In folklore, his mother supposedly dipped the young Achilles into a Yarrow bath to make him invincible. Since she held him by his heels, he was made vulnerable through his "Achilles heel."

Ecology & Ethnobotany: Yarrow is often referred to as "poor man's pepper." The leaves can be dried, ground, and used as seasoning. The young leaves can be added to salads. The aromatic leaves were also placed in freshly split fish to expedite drying.

Medicinally, the leaves and stems can be dried, boiled in water, strained and drunk to remedy a run-down condition or help with an upset stomach. Taken as a hot infusion, Yarrow will increase body temperature, open skin pores, and stimulate perspiration, making it a valuable herb for colds and fevers. The juice can be used as an eyewash to reduce redness. Leaves can be used to stop bleeding in small wounds, and to heal rashes when applied directly to the skin. Leaves were also chewed to relieve toothaches. A poultice of mashed leaves can be applied to swellings or sores. To date, over 100 biologically active compounds have been identified from the species (Foster and Duke 1990); some are known to be quite toxic.

Prolonged use of Yarrow may cause allergic rashes and make the skin more sensitive to sunlight.

Rubbing the plant on one's clothing and skin, was an ancient prescription for repelling biting insects. The stalks burned on coals were said to deter mosquitoes. The leaves were used in herbal snuffs and smoking tobaccos. Yarrow has also been used as a hops substitute for brewing Yarrow beer.

False or Mountain Dandelion (*Agoseris*)

Description: These are annual or perennial, tap-rooted herbs with milky juice that resemble Dandelion (*Taraxacum*). The flowers are all ray, yellow or occasionally orange in color. The pappus is white with hair-like bristles. The fruit (achene) is conspicuously ten-nerved. False Dandelions occur on moist to dry ground, in meadows and open areas at various elevations. The genus name is from the Greek, meaning "goat chicory." At least three species of Mountain Dandelion are known to occur in the southern California mountains.

"Quick Key" to the Mountain Dandelions

1. Heads less than 1 inch high - ***A. heterophylla***
1. Heads 1 to 1½ inches high - **2**

2. Leaf segments pinnatifid into regular lobes which point backwards; pappus 9/16 - 11/16-inch long; ligules $\frac{1}{4}$ - 9/16-inch long - *A. retrorsa*

2. Leaves irregularly lobed; pappus bristles 5/16 - $\frac{1}{2}$-inch long; ligules 1/8 - $\frac{1}{4}$-inch long - *A. grandiflora*

Agoseris (*A. grandiflora*) This Agoseris grows on dry or moist areas below 6,200 feet, and blooms from May to July.

Agoseris (*A. heterophylla*) This Agoseris grows on open, grassy slopes and flats below 7,500 feet. Blooms from April to July.

Agoseris (*A. retrorsa*) This Agoseris grows on dry ridges and slopes from 2,500 to 8,000 feet. Blooms from May to August.

Ecology & Ethnobotany: The leaves and roots of some species are edible when cooked but are bitter, especially in late season. Strike (1994) indicates that the seeds were eaten by the Chuvash Indians in southern California. The sap from the leaves of some species, when hardened can be used as chewing gum. Since the sap from some species is very thick and insoluble, it may be useful for waterproofing containers (e.g., coiled baskets) and footwear.

Ragweed (*Ambrosia*)

Description: In general, the 12 species of *Ambrosia* in California are annuals or shrubs with leaves that are opposite below and alternate above. The yellow flowers are arranged in spikes or racemes and the fruit is enclosed in a bur. There is no pappus. Since the wind-blown pollen is highly allergenic, Ragweeds are a notorious cause of hayfever where the plants are common. The genus name is from the Greek and refers to an early name for aromatic plants. It is also the mythic food of the gods.

Ecology & Ethnobotany: *Ambrosia trifida* (Great Ragweed) was cultivated in prehistoric times for its edible seeds in the midwestern United States. A tea from the leaves of *A. trifida* was formerly used for fevers, diarrhea, dysentery, nosebleeds, and gargled for sore throats. Other species were used in teas for various medicinal purposes. The heated leaves of *A. psilostachia* (Cuman Ragweed) were used as a poultice to ease aching joints, and a decoction was used to bathe bad sores and burns. Native Americans rubbed the leaves of *A. artemisiifolia* (Annual Ragweed) on insect bites, infected toes, minor skin eruptions, and hives. A tea was used for fevers, nausea, mucous discharges and intestinal cramping.

Western Pearly-everlasting (*Anaphalis margaritacea*)

Description: This perennial grows up to 36 inches tall and has white woolly herbage. Leaves are alternate, entire, lanceolate to linear or oblong in shape. The sessile leaves are also woolly beneath, but soon becoming green and glabrous above. The flowering heads form a flat topped, terminal cluster. The involucral bracts are papery, pearly white, and imbricated in several series. Pearly Everlasting is found in openings along trails and on talus slopes below 8,500 feet in the San Bernardino Mountains. Flowers from June to August.

Ecology & Ethnobotany: The herbage of Western Pearly Everlasting has been used as a tobacco substitute to relieve headaches. As a tea, the plant has been used for colds, bronchial coughs, and throat infections. The whole plant can be used as a wash or poultice for external wounds. It has also been used for rheumatism, burns, sores, bruises, and swellings.

Pussy-toes (*Antennaria*)

Description: In general, Pussy-toes are herbaceous often mat-forming perennials. The heads are discoid, with small, white flowers surrounded by bracts that are typically hairy below with a smooth and membranous portion varying in color from white to pink to dark brown or black. The pappus is composed of numerous hairy bristles. The species in southern California can be found in dry, open habitats, or in moist or seasonally wet places from the foothills to alpine areas.

"Quick Key" to the Pussy-toes

1. Heads solitary and terminal - *A. dimorpha*
1. Heads several to many - **2**

2. Basal leaves less hairy on upper surface than beneath
- *A. marginata*
2. Basal leaves hairy on both sides equally - **3**

3. Phyllaries blackish-green at tips - *A. media*
3. Phyllaries white or rosy colored at tips - *A. rosea*

Antennaria (*A. dimorpha*) This is a dwarf, matted perennial with silky, woolly herbage and grows on dry slopes and ridges from 4,700 to 9,700 feet on Mt. Pinos, and in the San Bernardino and San Gabriel Mountains. Flowers from May to July.

Pussy-toes (*A. marginata*) Found between 6,000 to 7,000 feet in forested areas of the San Bernardino mountains. Blooms from May to July.

Pussy-toes (*A. media*) Occurs above 11,000 feet in the San Bernardino mountains and northward. Flowers July to August.

Pussy-toes (*A. rosea*) This is a low perennial plant grows in the San Bernardino Mountains and northward from 4,500 to 12,000 feet. Flowers from June to August.

Ecology & Ethnobotany: The sap from the stem

of most species can be chewed like gum and has some nutritive value. Moore (1979) indicates that a tablespoon of the chopped plant steeped in hot water is an excellent remedy for liver inflammation. It has also been used as an astringent to the intestinal tract. Leaves can be made into a poultice for use on bruises, sprains, and swelling. The blossoms could be boiled and used to bathe sore or ulcerated feet, or mashed and applied to sores. A related species, *A. microphylla* (Littleleaf Pussytoes) was chewed as a cough remedy by the Thompson

Indians in British Columbia. The tiny leaves were also stripped, dried, and used as one of the ingredients in Indian tobacco.

Chamomile (*Anthemis*)

Description: Three species of *Anthemis* are found in California, *A. arvensis* (Corn Chamomile), *A. cotula* (Stinking Chamomile), and *A. tinctoria* (Golden Chamomile). They are annual or short-lived perennial herbaceous plants with radiate flowering heads composed of white or yellow ray flowers. Introduced European weeds, they are usually found at the lower elevations as escapees from gardens and cultivation.

Ecology & Ethnobotany: A tea can be made from Stinking Chamomile to induce sweating and vomiting. An astringent and diuretic, it has been used for ailments such as fevers, colds, diarrhea, dropsy, rheumatism, and headaches (Foster and Duke 1990). The leaves were rubbed on insect bites and stings. Golden Chamomile was originally considered a noxious weed of clover fields, but has since been brought into cultivation for horticultural purposes.

Common Burdock (*Arctium minus*)

Description: This is a coarse biennial with large rounded to ovate leaves. There are several flowering heads comprised of pink or purplish tubular flowers. This naturalized species from Europe is an occasional weed at the lower elevations. Flowers June to August. *Arctium lappa* (Greater Burdock) also occurs in California and can be used in much the same way as Common Burdock.

Ecology & Ethnobotany: Rich in vitamins and iron, the young leaves and shoots can be gathered for use as a potherb, or eaten raw in salad. The plant has a strong rank taste and an objectionable odor. The inner pith-like material of the young stems can be eaten raw, but we find it better when boiled in one or two changes of water. The roots of young plants can be sliced and cooked, then eaten. The older roots can be roasted and ground for use as a tea or coffee substitute. Seeds can be dampened and grown as sprouts.

The medicinal uses of the plant predate its use as a food plant. The Chinese are said to have used the plants as a blood purifier for thousands of years. Current research has confirmed the usefulness of Common Burdock in the treatment of rheumatism, water retention, and high blood pressure. As a wash, it was used externally for hives, eczema, and skin problems. The crushed seeds were used as a poultice.

The tall rigid stems were used as drills for primitive fire-starting techniques. The burs can be used as a survival "Velcro" for holding clothes together.

Arnica (*Arnica chamissonis*)

Description: This perennial plant grows up to 32 inches tall, and has herbage that is slightly hairy and glandular above. The leaves are opposite, lanceolate, and entire or slightly toothed. The flowering heads, numbering 5 to 15 have involucres up to $\frac{1}{2}$ inch long. Both ray and disk flowers are present. The ligules of the ray flowers are yellow and usually about 13 per head. The corollas of the disk flowers are 5/16-inch long, and yellow or orange. The pappus consist of many capillary bristles.

Arnica grows in meadows and in moist places from 5,000 to 11,000 feet. Flowers from July to August. The name translates as "lamb's skin" and refers to the modified leaves (bracts) that are usually woolly.

Ecology & Ethnobotany: All the species are reported to be poisonous if taken internally. Arnica contains arnicin, choline, a volatile oil, arnidendiol, angelic and formic acid, and other unidentified substances that can alter cardiovascular activity. The Federal Drug Administration listed *Arnica* as "unsafe" and bans its use for human consumption. Moore (1979) states that Arnica is an external remedy only. The chopped plant is steeped in rubbing alcohol for about a week and squeezed through a cloth. The liniment is then used for joint inflammations, sprains, and sore muscles. It should not be used if the skin is broken since it is toxic if it enters the bloodstream. Arnica is useful as a topical preparation for bruises, sprains and other closed injuries. When gathering, grasp the plant at the base of the stem just below the ground to leave the rhizome for continued growth. Wear gloves as the volatile oils can be absorbed.

Warning: All species of Arnica are reported to be poisonous if taken internally.

Sagebrush, Wormwood (*Artemisia*)

Description: There are a number of species of *Artemisia* in southern California, including annual, biennial, and perennial herbs and shrubs. They are mostly aromatic with entire or dissected leaves. The flower heads are small, inconspicuous, and comprised of disk flowers. The genus name honors Artemisia, wife of Mausolus who was the King of Caria (a province in Asia Minor). After the

King's death in 350 BC, Artemisia built the renowned Mausoleum, one of the Seven Wonders of the World.

Mugwort (*A. douglasiana*) Mugwort grows in low places up to 6,000 feet. Flowers from June to October.

Tarragon (*A. dracunculus*) Tarragon occurs on dry, disturbed places below 9,000 feet. Blooms from August to October.

Western Mugwort (*A. ludoviciana*) Western mugwort grows on dry, open places below 8,500 feet in the San Jacinto and San Bernardino mountains. Flowers from July to September. Also occurring in the area are two subspecies.

Sagebrush (*A. nova*) Occurs in dry places in sagebrush scrub and juniper woodlands in the San Jacinto mountains and desert slopes of the San Bernardino mountains and northward. Blooms September to November.

Sagebrush (*A. rothrockii*) This low, spreading, evergreen shrub grows on dry or wet, rocky slopes from 6,500 to 11,500 feet in the San Bernardino Mountains. Blooms from August to September.

Great Basin Sagebrush (*A. tridentata*) This evergreen shrub up to 9 feet tall, has silvery gray herbage and is quite common on dry slopes and plains on the desert side of the mountains, and occurs between 1,500 to 10,600 feet. Blooms from August to October.

Ecology & Ethnobotany: The seeds of many species are edible raw or as flour. The seeds and peeled shoots of *A. douglasiana* and *A. ludoviciana* were eaten raw by Native Americans in California.

Herbage of various *Artemisia* species may be toxic if eaten in large amounts, but may be used in small quantities to flavor stews, soups, and other foods. A tea from leaves was a cure for colds and sore eyes, and was used as a hair tonic. Some of the "softer" species can be used as toilet paper and foot deodorant. Crushed leaves can be mixed with stored meat to maintain a good odor. Since many species are aromatic, they can be used to store buried food caches by masking the odor of foodstuff, and to rub on the body to mask human scent while hunting. The wood of *A. tridentata* is a good material for fire drills. Although cordage can be made from the bark, it is not very strong.

Many species of Artemisia have been used as medicine by Native Americans and were used in sweathouses to relieve numerous ailments. The bitter leaves of *A. absinthium* (Absinth Sagewort) can be nibbled on to stimulate an appetite. A strong tea of *A. ludoviciana* was used as an astringent for eczema, and as a deodorant and anti-perspirant for underarms and feet. A weak tea was used for stomachaches. For sinus ailments, headaches, and nosebleeds, a leaf snuff was used. A leaf or root tea of *A. dracunculus* was used for colds, dysentery, headaches, and to promote an appetite. The leaves were made into a poultice and used on wounds and bruises. Moore (1979) says that *A. tridentata* is strongly antimicrobial and was used as a disinfectant and cleansing wash. Volatile oils in *A. tridentata* are responsible for its pungent aroma and are so flammable that they can cause even green plants to burn. It should also be noted that the Federal Drug Administration classifies Artemisia as an unsafe herb containing "...a volatile oil which is an active narcotic poison" (Duke 1985).

Wilderness Food Storage Pit

To store foods for extended periods of time, Native Americans used storage pits. After digging a hole, moisture was removed from the soil by lining the pit with hot rocks and allowing it to steam. With the rocks left in place, the pit was then lined with dried grasses, and food was placed inside. On top of the food, dried bark from junipers or other plant high in tannic acid were placed to repel insects. On top of this, dried, aromatic, non-poisonous leaves such as sagebrush were placed to disguise the smell of the food. Lastly, the pit was covered with a thick layer of dirt and heavy rocks to prevent animals from uncovering the food stored.

Aster (*Aster*)

Description: The many Aster species in southern California are perennial herbs with alternate leaves. The ray flowers are pistillate containing female parts, ranging in color from blue, purple, white to pink (never yellow). The central disk flowers are usually yellow. The involucral bracts are in many overlapping series, like shingles on a roof. Pappus is composed of capillary bristles. Asters usually flower from late summer into fall. A related genus, *Erigeron* (Fleabane) is often confused with Aster, but Fleabanes usually flower from late spring to mid-summer and the bracts of the involucre are in 1-2 series. Asters are found in various habitats from low elevations into the alpine zone. The genus name is from the Greek and Latin meaning star, referring to the radiating ray flowers.

Aster (*A. alpigenus*) This Aster grows in moist or boggy meadows from 4,000 to 11,500 feet on Mt. San Jacinto and northward. Blooms from June to September.

Long-leaved Aster (*A. ascendens*) This Aster grows in moist or dry soil from 200 to 7,500 feet in the San Gabriel, San Jacinto, and San Bernardino Mountains. Blooms from July to October.

Alpine Leafy-bract Aster (*A. foliaceus*) This species occurs in open woods to alpine meadows in the San Gabriel and San Jacinto mountains between 4,500 to 10,000 feet.

Aster (*A. occidentalis*) This Aster grows in moist meadows from 4,000 to 10,500 feet. Blooms from July to September.

Ecology & Ethnobotany: The leaves of several species were boiled and eaten by some Native American tribes. Some Asters are known to absorb selenium. The Cheyenne Indians used a tea from *A. foliaceus* in the form of eardrops to relieve earaches. *Aster laevis* (Smooth Aster) was burned to create smoke in a sweatbath, and the crushed foliage of *A. hesperius* (=*A. lanceolatus* ssp. *hesperius*) (Siskiyou Aster) was sprinkled on live coals and inhaled to treat nosebleeds.

Balsamroot (*Balsamorhiza*)

Description: These are low perennial herbs with thick rootstalks, and the leaves are mostly basal, large, and long-petioled. The yellow flowering heads are large showy, mostly on long peduncles. Balsamroot is often confused with *Wyethia* (Mule's Ears), which can be found in similar habitats. However, *Wyethia* leaves lack the fuzzy gray appearance seen on the Balsamroot.

Balsamroot (*B. deltoidea*) This is a perennial herb can be found in open areas from 2,000 to 7,000 feet.

Balsamroot (*B. sagittata*) This perennial plant grows from 4,300 to 8,300 feet in Kern and Inyo Counties and northward. It is common in sandy soil. Flowers from May to July. Interestingly, a small patch can be seen on one of the ski runs near Big Bear.

Ecology & Ethnobotany: Although Balsamroot (*B. sagittata*) is considered one of the most versatile sources of food, it is not necessarily palatable. The plants contain a bitter, strongly pine-scented sap. The large taproot, root crowns, young shoots, young leafstalks and leaves, flower budstalks, and the seeds were all eaten by various Native Americans. The larger mature leaves were often used in food preparation (i.e., wrapping).

The woody taproot of perhaps all species is edible raw or cooked. The polysaccharide inulin is the major carbohydrate found within the root. The roots can be collected throughout the year, but are very difficult to dig out. In some species, the taproot may be as large as one's forearm. Cooking the roots is yet another challenge. One method we have used involves peeling the roots by pounding them to remove the bark. These were then pit cooked for 24 or more hours. When properly cooked, the roots turn brownish and sweet tasting. Another way to prepare the roots is to pit steam large quantities for a day and then mash and shape them into cakes for storage. Cooked this way, the roots were called "pash" or "kayoum."

The young shoots are edible raw or pit cooked before they emerge in early spring. The young stems and leaves can also be eaten raw or boiled as greens. The older stems are fibrous, tough, and will require some additional boiling.

The flower budstalks are collected while the buds are still tightly closed, then peeled and eaten raw or cooked as a green vegetable. They have a slightly nutty taste.

When harvested from dried heads, the seeds can be roasted and eaten or ground into flour. The chaff is usually removed by winnowing.

The roots are said to be antimicrobial, expectorant, disinfectant, and immuno-stimulant. They can be mashed and applied to swellings and insect bites. Native Americans considered a boiled solution from the root of *B. hirsuta* (= *B. hookeri* var. *neglecta*) (Neglected Balsamroot) root to be an excellent medicine for stomach aches and bladder troubles. The mashed roots of Arrowleaf Balsamroot were also used by Native Americans to treat swellings or insect bites.

Brickellbush (*Brickellia*)

Description: The three species of *Brickellia* that occur in southern California are perennial herbs with fibrous roots. The disk flowers are all tubular, white or creamy to pink-purple. They can be found in a variety of habitats at the lower elevations. This is a large and complex genus consisting mostly of shrubs.

California Brickellbush (*B. californica*) California Brickellbush is found in washes and on dry slopes of the foothills and chaparral, usually below 8,000 feet. Blooms from August to October.

Little-leaved Brickellbush (*B. microphylla*) Little leaved Brickellbush is found on dry, rocky places from 3,000 to 8,000 feet in the San Gabriel Mountains and the west Mojave Desert. Blooms from August to November.

Brickellbush (*B. nevinii*) This *Brickellia* grows on dry slopes and washes between 1,000 to 5,500 feet in chaparral areas, from Kern County southward. Blooms from September to November.

Ecology & Ethnobotany: Moore (1989) states that a tea or tincture from *B. grandiflora* (Tasselflower

Brickellbush) has three distinct uses: 1) lowering blood sugar in certain types of diabetes; 2) stimulating hydrochloric acid secretions by the stomach; and 3) stimulating bile synthesis and gallbladder evacuation. Others species were also probably used medicinally by Native Americans.

Tinctures

Here is one general procedure for making tinctures. First, place the fresh plant in a glass jar and cover with 80 proof brandy or vodka. Keep the jar in a warm dark place for about two weeks, shaking it daily. After two weeks, strain the herbs through muslin or two layers of cheesecloth. Squeeze well to extract as much fluid as possible. Discard the herbs and bottle the tincture. The dosage varies depending on what the herb is to be used for. For Sundew, 3-6 drops in a cup of water is recommended. Vinegar can be used in place of alcohol, but has a shorter shelf life (1-2 years) than alcohol tinctures (30-40 years). Vinegar tinctures are often used for babies, alcoholics, and persons with liver problems.

Plumeless Thistle (*Carduus*)

Description: Three species are reported to occur in southern California: *C. nutans* (Nodding Plumeless Thistle), *C. pycnocephalus* (Italian Plumeless Thistle), and *C. tenuiflorus*. *Carduus* is distinguished from *Cirsium* in that the pappus of *Carduus* is simple and smooth, not a plume. These are weedy species that may occasionally be found along roadsides and other waste places at the lower elevations.

Ecology & Ethnobotany: Kirk (1975) indicates that the pith of four species (unspecified), without the easily removed rind, may be boiled in salted water and seasoned in various ways. The dried flowers may be used as a rennet to curdle milk. Additionally, Strike (1994) indicates that the raw or cooked leaves and stems, and raw buds were also eaten.

Pincushion, Chaenactis (*Chaenactis*)

Description: These are biennial or perennial herbs from a taproot. The leaves are pinnately dissected and the flowering heads are comprised of disk flowers which are white to pink to rose in color. The six species in California can be found in open, dry and rocky habitats

from the lower elevations into the alpine. The genus is endemic to the western United States.

Chaenactis (*C. glabriuscula*) This is a low growing annual grows on sandy or gravelly areas up to 6,500 feet. Flowers from April to June.

Pincushion Flower (*C. santolinoides*) This slightly matted perennial plant with glandular, mostly leafless stems grows in open pine woods and on dry ridges from 4,500 to 8,000 feet on Mt. Pinos and in the San Bernardino Mountains. Flowers from June to July.

Chaenactis (C. xantiana) This stout annual is common on slopes of desert mountains from 1,400 to 7,000 feet. Flowers from April to June.

Ecology & Ethnobotany: Leaves of *C. douglasii* (Douglas' Pincushion) were mashed and used to poultice sprains and swellings. A decoction of the plant was used for indigestion, coughs, and colds. Mashed leaves used on rattlesnake bites.

Chicory (*Cichorium intybus*)

Description: This is a perennial herb that grows up to three feet tall with dandelion-like leaves. The blue flower heads, which can be seen from spring to fall, are composed of 15-20 or more ray flowers. The sap is milky. Chicory is a plant of waste places and is found at the lower elevations. Introduced from Europe, it now grows throughout the United States.

Ecology & Ethnobotany: While the roasted root was used for coffee, it is not considered a very satisfactory substitute by itself. Many coffee producers have used Chicory as a coffee additive.

The young basal leaves and flowers buds hidden at the base of the leaves are edible and best if collected

from fall to spring. Because they are bitter, we found it necessary to boil them in at least 1 to 3 changes of water. When collected very young, the plants are milder when eaten raw. In some European countries, the buds are pickled and canned (Tull 1987).

Chicory Coffee Additive

To make a coffee additive, dig up Chicory roots in the fall through spring, scrub them, and slice in half. Roast them in an oven at a low temperature (e.g., 250 degrees Fahrenheit) for 2 to 4 hours, or when they become dark brown and brittle. Break up and grind as you would coffee. One-part Chicory to 4 parts coffee is a common ratio when brewing.

Rabbitbrush (*Chrysothamnus*)

Description: The three species of *Chrysothamnus* in southern California are shrubs with alternate, sessile, entire, and linear leaves. The flowering heads are comprised of 5-30 yellow disk flowers and typically bloom in late summer and fall. Rabbitbrush is found in dry, open places at low to middle elevations.

"Quick Key" to the Rabbitbrush

1. Heads in a leafy raceme - *C. parryi*
1. Heads in rounded, terminal clusters - **2**

2. Involucres 3/8 - $\frac{1}{2}$-inch high; stems with gray or white felty pubescence; ill-smelling shrub - *C. nauseosus*
2. Involucres 5/16-inch high or less; stems glabrous and glandular - *C. viscidiflorus*

Common Rabbitbrush (*C. nauseosus*) This ill smelling shrub grows in dry areas from 6,000 to 9,500 feet in the San Gabriel, San Bernardino, and San Jacinto Mountains. Flowers from July to September.

Rabbitbrush (*C. parryi*) This is a leafy shrub blooms from July to September. Two subspecies also occur in the area.

Sticky-leaved Rabbitbrush (*C. viscidiflorus*) This shrub is common in dry, open areas from 4,000 to 7,500 feet in the San Jacinto and San Bernardino Mountains. Flowers from July to September.

Ecology & Ethnobotany: A tea was reported to be made from the twigs of *C. nauseosus* that provided relief from chest pains, coughs, and toothaches.   The leaves and stems were also boiled and the liquid was used to wash itchy areas.

Great Basin Indians were accustomed to chewing the stems of Rabbitbrush to extract the latex.

They believed that chewing Rabbitbrush relieved both hunger and thirst. The secretion obtained from the top of the roots can also be chewed as gum.

The rubber shortage of World War II stimulated research on Rabbitbrush and other rubber-producing plants. Rabbitbrush produces a high-quality rubber called chrysil that vulcanizes easily. Extraction of this rubber for economic reasons at this point is not feasible. Because of their rubber-based compound, Rabbitbrush will burn even if it is wet or green. Navajo Indians derived a yellow dye from the flowers, while the inner bark yielded a green dye.

Thistle (*Cirsium*)

Description: The many species of Thistle that occur in southern California are characterized as biennial or perennial herbs with alternate leaves that are lobed or cleft with spines. The red, yellow, or white heads are showy and the involucral bracts are overlapping. The native and introduced species can be found in a wide variety of habitats from the foothills to the higher elevations. *Cirsium* comes from the Greek kirsos, meaning "swollen vein", for which thistles (kirsios) were a reputed remedy. At least five species of Thistle can be found in the mountains of southern California

Canada Thistle (*C. arvense*) This species is a native of Europe and is found in disturbed areas below 6,000 feet in the southern California mountains. It is considered to be a noxious weed.

Western Thistle (*C. occidentale*) This slender and spiny biennial is a common plant on dry slopes below 10,000 feet. Blooms from April to July.

Yellowspine Thistle (*C. ochrocentrum*) This species is found in disturbed areas below 5,500 feet in the Transverse and Peninsular Ranges and is a native to central United States. It is also considered a noxious weed.

Elk Thistle (*C. scariosum*) This species of thistle is found in moist places up to 10,000 feet in the San Gabriel, San Bernardino, and Peninsular mountain ranges. It is extremely variable with tall and dwarf plants sometimes occurring together. Blooms June to August.

Bull Thistle (*C. vulgare*) This common species is usually found in disturbed areas up to 7,000 feet in the southern California mountains. It is a native of Europe.

Ecology & Ethnobotany: Thistles were not a major food source in the past, but were used when needed. Truman Everts, a participant in the early explorations of the Yellowstone Park region, became lost for more than a month and subsisted on thistles. He apparently had lost his glasses and was able to identify Thistles by touch. Although Thistles are difficult to collect, they are well worth the pain.

All species have roots that can be eaten raw, boiled or roasted. Some have roots that turn sweet when roasted. The immature flower buds (asparagus-like) can be eaten raw or cooked. Young leaves de-thorned are edible raw, and a tea can be brewed from all leaves. The peeled young stems may be cooked as greens, and resemble celery in taste. The older stalks are also edible but are somewhat more fibrous and bitter. The seeds can be boiled and eaten in the same manner as sunflower seeds, or they can be ground into flour for baking.

Medicinally, Thistle stalks were chewed to ease stomach pains. Pounded stalks were used as a salve for facial sores or on infected wounds. A decoction made from Thistle roots was used to relieve asthma.

When well dried and de-thorned, stems can be used as hand drills for starting fires. The stem fibers of any thistle species can be used as thread or crude cordage. To obtain the fiber, simply soak the stalks in water for a day or more to loosen them from the outer layer. The downy part of seed heads makes good insulating material and a good tinder additive.

Canadian Horseweed (*Conyza canadensis*)

Description: This is an annual weed similar to Erigeron that grows to about two feet tall with numerous, narrow leaves. There are numerous white flower heads. Canadian Horseweed is usually found growing in waste places at the lower elevations, below 6,000 feet.

Ecology & Ethnobotany: A native to North America, Canadian Horseweed was introduced into Europe around the mid-17th century where it became widely known for its tonic and astringent properties. A tea was made from the entire dried plant and used for gravel dropsy, diarrhea, and scalding urine. Native Americans used the plant in the form of a tea for leucorrhea, and applied the solution to external sores in cases of gonorrhea (Callegari and Durand 1977). Foster and Duke (1990) also indicate that *E. canadensis* (= *C. canadensis*) was used as a folk diuretic, astringent for diarrhea, kidney stones, nosebleeds, fevers, and cough. The leaves and tops of Horseweed can be pounded and eaten uncooked (Strike 1994).

Corethrogyne (*Corethrogyne filaginifolia*)

Description: This is a perennial herb clothed with a soft white deciduous wool, resembling an Aster. The genus name is from the Greek, *korethron*, a brush for sweeping and *gyne*, for style, from the brush-like style appendages. There are about 11 varieties of this species. Flower June to October.

Ecology & Ethnobotany: Native Americans in Northern California made an infusion of leaves and twigs which then was burned in sweathouses to induce sweating. Inhaling the herbal steam was said to relieve colds. Additionally, an infusion of boiled flowers was drunk to relieve chest pains.

Hawksbeard (*Crepis*)

Description: In general, these are perennial, tap-rooted herbs with milky juice. The leaves are alternate or all basal, and the yellow flowers are all ray. The various species can be found in dry open places at lower elevations to gravelly or rocky places in alpine or subalpine areas.

Long-leaved Hawksbeard (*C. acuminata*) This perennial plant grows in dry places from 2,600 to 11,000 feet in the San Bernardino Mountains and north. Flowers from June to August.

Low Hawksbeard (*C. modocensis*) This perennial grows from 5,000 to 10,000 feet in the San Bernardino Mountains. Flowers from June to July.

Dwarf Hawksbeard (*C. nana*) This is a tufted perennial with slender stems grows on stony or gravelly talus from 8,000 to 13,000 feet in the San Gabriel Mountains and north. Flowers from July to August.

Western Hawksbeard (*C. occidentalis*) This is a perennial plant that grows in Kern and Ventura Counties and northward from 4,000 to 9,000 feet. Flowers from June to August.

Ecology & Ethnobotany: The stems and leaves of *Crepis* were eaten by Native Americans. The Karok Indians of Northern California peeled the stems of *C. acuminata* before eating.

The seeds or whole plant of *C. acuminata* was thoroughly crushed and applied as a poultice to breasts after childbirth to induce milk flow. The root of the plant was used to remove a foreign object from the eye. The root can also be ground into a smooth powder and sprinkled in the eye to treat eye problems. Several applications were necessary.

Mountain Bush Sunflower (*Encelia virginensis*)

Description: This low shrub grows up to 60 inches tall. The leaves are alternate, lanceolate oblong to ovate, green and pubescent. The flowering heads are solitary, and occur at the ends of the stem. The pappus consists of 2 awns or none. *Encelia* grows in gravelly areas in canyons and on the desert side of the mountains between 1,000 to 5,000 feet. Flowers from April to May.

Ecology & Ethnobotany: The leaves, roots, and flowers of Mountain Bush Sunflower were used to produce a wash for relieving rheumatic pain.

The heated gum of a related species, *E. farinosa* (Brittlebush), was applied externally to relieve chest pain. Additionally, the flowers, leaves, and stems were made into a decoction that, when held in the mouth, relieved toothaches.

Fleabane (*Erigeron*)

Description: Many species of Fleabane occur in California. They are characterized as annual, biennial, or perennial herbs with alternate or basal leaves. The flowering heads are radiate with narrow ray flowers that may be white, pink, blue, purple, or occasionally yellow. The numerous disk flowers are yellow, and the pappus is comprised of capillary bristles. The various species bloom mostly in the spring and early summer, except at the higher elevations (see Aster), and can be found in a variety of habitats. Fleabanes resemble Aster, but are distinguished by fewer rows of involucral bracts. Fleabanes also bloom earlier than Asters. The genus name comes from the Greek *eri* (early) and *geron* (old man). The common name, Fleabane, comes from the belief that these plants repelled fleas.

Erigeron (*E. breweri*) This perennial plant grows on dry, rocky places in the San Gabriel and San Bernardino Mountains from 5,000 to 10,500 feet. Flowers from July to August.

Cut Leaf Daisy (*E. compositus*) This is a dwarf, cushion like plant that grows on rocky slopes from 8,000 to 13,000 feet on Mt. San Gorgonio. Flowers from July to August.

Diffuse Daisy (*E. divergens*) This is a branched biennial that grows in sandy soil in the San Gabriel, San Bernardino, and Laguna Mountains, between 200 to 7,500 feet. Blooms from March to November.

Leafy Daisy (*E. foliosus*) This is an erect perennial that grows on grassy or brushy slopes up to 5,500 feet. Blooms from May to July.

Parish's Erigeron (*E. parishii*) This is a perennial plant with a woody root and erect stems grows on dry slopes from 3,500 to 5,000 feet on the north face of the San Bernardino Mountains. Blooms from May to June.

Ecology & Ethnobotany: Fleabanes, in general, are listed as astringent and diuretic. The disk flowers of *E. philadelphicus* (Philadelphia Fleabane) were powdered to make a snuff to cause one to sneeze and breakup a cold or catarrh. A tea from the entire plant of *E. annuus* (Eastern Daisy Fleabane) was used to treat a sore mouth. The dried roots, stems, and flowers of *E. peregrinus* (Subalpine Fleabane) were steeped in hot water and the patient would breathe the vapors. Fleabanes may cause dermatitis in some people.

Woolly Sunflower (*Eriophyllum*)

Description: These are hairy annual to perennial plants that are herbaceous to shrubby. The genus name is from the Greek, *erion*, meaning wool, and *phyllon* for leaf.

"Quick Key" to the Woolly Sunflowers

1. Ray flowers 4-6 per head; ligules 1/8 - 3/16-inch long -
E. confertiflorum
1. Ray flowers 8-13 per head; ligules 3/8 - ¾-inch long -
E. lanatum

      Golden Yarrow (*E. confertiflorum*) This perennial plant is common on brushy slopes to 8,000 feet. Flowers from April to August.
      Eriophyllum (*E. lanatum*) This is a herbaceous perennial is common on brushy slopes up to 10,000 feet. Flowers in June and July.

    Ecology & Ethnobotany: *Eriophyllum* seeds were parched and ground into a flour by Cahuilla and Luiseno Indians in California. The seeds were also incorporated into pinole.

Cudweed (*Gnaphalium*)

    Description: These are woolly annual, biennial, or short-lived perennials herbs with alternate leaves. The plants are often confused with *Antennaria*, but Cudweeds have both male and female flowers on the same plant and are tap-rooted. The disk flowers are yellow or whitish. The species are found from the low to mid elevations in moist, open areas to well-drained soils.

"Quick Key" to the Cudweeds

1. Plant 8-36 inches tall; heads ¼-inch high, not surpassed by the subtending leaves - *G. stramineum*

1. Plant mostly 2-8 inches high; heads small 1/8-inch high, immersed in wool and surpassed by the subtending leaves - *G. palustre*

> **Lowland Cudweed (*G. palustre*)** This odorless annual is found below 9,500 feet. Flowers from May to October
> **Gnaphalium (*G. stramineum*)** This is an aromatic annual or biennial plant that grows in moist, open, waste places below 6,000 feet. Flowers from June to October.

Ecology & Ethnobotany: The bruised plant assists in healing wounds, and steeping the leaves in cold water is used for increasing perspiration. Some species contain pyrrolizidine alkaloids and should be regarded as potentially toxic.

Gumweed (*Grindelia*)

> **Description**: These species are biennial or short-lived perennial of waste places at low elevations. The leaves are alternate and have toothed margins. A sticky, resinous sap covers the leaves and bracts of the yellow flowers.
> **Ecology & Ethnobotany**: In general, Gumweeds are considered toxic, and the toxicity appears to be dependent upon the soil in which it grows. However, many species have been used medicinally for hundreds of years.
> The sticky flowers heads of *G. squarrosa* (Curlycup Gumweed) were used as a chewing gum substitute. The young leaves make an aromatic, bitter tea. The flower heads can be boiled in water and used as

an external remedy for skin diseases, scabs, and sores. A hot poultice of the plant was used for swellings. A tea made from the plant was used for coughs, pneumonia, bronchitis, asthma, and colds.

Caution: *G. squarrosa* tends to concentrate selenium and is therefore considered toxic.

Broom Snakeweed (*Gutierrezia sarothrae*)

Description: This is a sticky, glandular perennial, appearing almost shrubby, and grows 6 to 24 inches high. The leaves are entire, linear filiform, and less than 1/8-inch wide. The flowering heads are small, numerous, and with whitish, leathery, involucral bracts. Ray flowers are yellow, number 3 to 8 per head, and are 1/8 inch long. There are 3 to 8 disk flowers, and the pappus consist of 2 to 8, stiff awns. Broom Snakeweed grows on dry slopes up to 8,000 feet. Blooms from May to October. Another common name, Matchweed, refers to the match-like appearance of the flower heads.

Ecology & Ethnobotany: As with many aromatic plants, this species was used medicinally. The plant was boiled to make a tea for colds, coughs and dizziness. The tops of fresh, mature Snakeweed were boiled until strong and dark. The liquid could be drunk for lung trouble and colds, or applied externally for skin ailments such as heat rash, poisoning, and athlete's foot. For respiratory ailments, the root was boiled in water and the steam inhaled.

Haplopappus (*Haplopappus*)

Description: These are herbs or shrubs that are often glandular. The leaves are alternate, entire to pinnatifid, and often thick. Many members of this genus are now considered to be *Ericameria*

"Quick Key" to the Haplopappus

1. Flowers not with both ray and disk flowers; either ray alone or disk - **2**
1. Both ray and disk flowers present - **3**

2. Leaves ovate to spatulate - *H. cuneatus*
2. Leaves linear to lanceolate - *H. parishii*

3. Leaves filiform, round in cross-section (pine-like) - *H. pinifolius*
3. Leaves flat and linear - *H. linearifolius*

Haplopappus (*H. cuneatus = E. cuneata*) This shrub grows in rock crevices between 2,500 to 9,000 feet. Blooms from September to November.

Haplopappus (*H. linearifolius = E. linerifolia*) This shrub grows on dry slopes and rocky or sandy areas below 6,000 feet. Flowers from March to May.

Haplopappus (*H. parishii = E.p*) This shrub is found in the San Gabriel, San Bernardino, San Jacinto, and Santa Ana Mountains, and south through San Diego. Flowers from July to October.

Pine-bush Haplopappus (*H. pinifolius = E. pinifolia*) This shrub is common on dry slopes from Los Angeles County to San Diego County. Flowers In the spring from April to July and in the fall from September to January.

Ecology & Ethnobotany: The seeds and stems of Haplopappus were eaten by many Native Americans. A decoction from the stem of *H. cuneatus* was used to treat colds.

Sneezeweed (*Helenium bigelovii*)

Description: This stout biennial grows up to 32 inches high, and the lower leaves are linear lanceolate to lanceolate. The flowering heads are solitary and showy, and occur on long, leafless peduncles. Both ray and disk flowers are present. The disk flowers are globose, and yellowish brown in color. The ray flowers, numbering about 13 to 30 per head, are with yellow ligules with 3 lobed apex. The ligules are turned downward, giving the disk flowers a globose and prominent appearance. Sneezeweed is common in moist, meadowy places from 5,000 to 8,500 feet. Flowers from June through August. The plant is so named because the appearance of the rays suggests that it has just sneezed.

Ecology & Ethnobotany: Powdered Sneezeweed was used as snuff to induce sweating, which in turn relieved the congestion of head colds.

Sunflower (*Helianthus*)

Description: These are coarse, annual and perennial herbs, often with tall stems. Leaves are simple, the lower ones are opposite, other sometimes alternate. Flower heads are showy with bright yellow ray flowers.

Involucral bracts are green and herbaceous. Other genera, including *Wyethia*, *Balsamorhiza*, and *Arnica*, are often mistaken for *Helianthus*. The genus name comes from the Greek *helios anthes*, which means sunflower.

Sunflower (*H. annuus*) This plant graces roadsides and other disturbed areas up to 5,000 feet, flowering from late spring to early fall. This is the ancestor of numerous commercial field and garden cultivars.

Wild Mountain Sunflower (*H. gracilentus*) This erect perennial grows on dry hillsides from 200 to 6,000 feet. Flowers from May to October.

Nuttall's Sunflower (*H. nuttallii*) This species occurs in damp meadows and marshes up to 7,500 feet in the Transverse and Peninsular Ranges.

Ecology & Ethnobotany: The largest member of this genus is *H. annuus* (Common Sunflower), is a valuable and useful plant. It has been cultivated in the United States since before Columbus. Other species of Helianthus may be used similarly.

The seeds may be eaten raw or roasted, then ground into meal and made into bread. The roasted shells can be used as a coffee substitute. To separate large amounts of seeds from shells, first grind them coarsely, then stir vigorously in water. In this way the shells will float, while the seeds sink to the bottom. The tiny unopened flower buds are also edible with a flavor similar to artichokes. To reduce their bitterness boil in 2-3 changes of water. Serve with lemon and melted butter.

Sunflower oil can be extracted from the seeds for cooking, and can also be used in making soap, paints, varnishes, and candles. It is extracted by simply boiling the crushed seeds and then skimming the oil from the surface of the water. The pulp remaining after the oil is extracted also provides food for livestock.

Medicinally, the crushed roots can be applied to bruises. Other uses of Sunflower include fiber obtained from the stalks for cordage, weaving, and sewing. The Chinese reportedly use the stalk fibers in fabrics and the pulp for paper production; the Russians used the stalks as buoyant material for life preservers. Purple and black dyes can be obtained from the seeds and a yellow dye from the flowers.

Butterflies and Edible Plants

From an ecological perspective, all life ultimately depends on other forms of life to survive and reproduce. As John Muir (1838-1914) once said, "When we try to pick out anything by itself, we find it hitched to everything else in the universe."

There are numerous ways to locate edible and useful plants. One means includes actively searching for the plants. This direct approach requires knowledge of the plants' basic life history, ecology, and distribution. Another way to locate potential useful plants, though indirect, but requiring knowledge of the plants is watching butterflies (Order Lepidoptera).

Most butterflies are closely tied to a specific species or group of plants to complete their life cycle. Specifically, the larvae (caterpillars) fed on those host plants. Adult female butterflies select the proper larval food plant by "smelling" with their antennae and "tasting" with sensory receptors on their feet. Eggs are then only laid on a specific host plant, so that hatching caterpillars can begin to feed right away. Here are some butterflies and their host plant(s) and uses:

Western White (Pontia occidentalis) are usually associated with Mustards (Brassicaceae). Mustards such as Lepidium, Brassica, Arabis, and Sisymbrium are edible as greens.

Milbert's Tortoiseshell (Nymphalis milberti) is associated with nettle (Urtica). These plants provide useful fibers for cordage, and the plants are edible when prepared as potherbs.

Fritillary (Speyeria) are found on violets (Viola) which are edible raw or mixed with other greens in a salad.

California Crescent (Phyciodes oreis) associates with thistles (Cirsium). These plants are edible raw or cooked, the stems can be used in hand drill fire starting, and fibers from the stem can be used as a crude cordage.

Hawkweed (*Hieracium*)

Description: Hawkweeds are fibrous rooted perennial herbs with milky juice. Flowers are all ray, yellow to sometimes orange or white in color. The name "hawkweed" comes from the belief by the ancient Greeks that hawks would tear apart a plant called the *hieracion* (from the Greek *hierax* meaning 'hawk') and wet their eyes with the juice to clear their eyesight. Hawkweeds are found in a variety of habitats up to the subalpine.

The two species in the southern California mountains include:

White-flowered Hawkweed (*H. albiflorum*)
This is an erect perennial plant with white flowers that grows in dry, open, wooded places below 9,700 feet. Flowers from June to August.

Hawkweed (*H. horridum*) This perennial plant with yellow flowers is common on dry and rocky slopes from 5,00 to 11,000 feet in the Santa Rosa and the San Jacinto Mountains and northward. Blooms from July to August.

Ecology & Ethnobotany: The green plant and juices of White-flowered Hawkweed may be used as a substitute for chewing gum, although it is best when dried first. The plant was also used to ease toothaches, to cure warts, or as an astringent in treating hemorrhages, and as a general tonic.

Wild Lettuce (*Lactuca serriola*)

Description: This is a tall, prickly plant with alternate leaves and milky juice. The yellow, blue, or whitish flowers are all ray. The pappus is white to brownish. This is a rather common weed in fields and waste places in the lower elevations.

Ecology & Ethnobotany: Collected in the late fall to early spring, the plants should be boiled in a couple of changes of water to reduce the bitterness. The earlier or younger the plant is collected, the better the flavor. Because of the latex sap, raw greens can cause upset stomach if eaten in quantity. In sensitive people, the latex can cause dermatitis. These wild plants contain more Vitamin A than spinach and a good quantity of Vitamin C. An extract of the white sap from two species of *Lactuca* in Europe has been used to replace opium in cough remedies. The extract, lactucarium, is reported to be a mild sedative. The plants also contain a mildly narcotic compound in the latex. The active constituents increase during flowering and are relatively low in young plants.

White Layia (*Layia glandulosa*)

Description: This annual grows up to 24 inches tall and has leaves that are rough hairy, linear to lanceolate in shape, with the basal ones, being toothed or lobed while the upper ones are entire. The flowering heads have both ray and disk flowers present. There are about 25 to 100 disk flowers. *Layia* is found in sandy soil up to 7,800 feet. Blooms from March to June.

Ecology & Ethnobotany: The seeds of *Layia* were often used in making pinole. The seeds of this species are edible after grinding it into flour for mush.

Lessingia (*Lessingia glandulifera*)

Description: This erect and branched annual plant grows 4 to 24 inches tall, and has stems that are woolly at the base and glabrous above. The leaves are alternate, with the basal ones narrow oblanceolate and pinnately cleft into 4 to 5 pairs of oblong lobes. The main leaves are reduced in size, sessile, toothed or entire. The yellow flowering heads have 25 to 38 disk flowers. The pappus consists of 21 to 31, brownish, capillary bristles. Lessingia grows on open areas, pine forests, and sandy soils up to 5,500 feet. Flowers from May to November.

Ecology & Ethnobotany: This species was used medicinally by some Native Americans. For example, the bark provided a treatment for general aches and pains. In the late summer, the outer bark was peeled off the stems and rolled into small pellets. A few pellets were then placed on the body where there was pain and lit on fire. When these pellets were completely burned, they were replaced by others, and the process repeated.

Skeletonplant (*Lygodesmia*)

Description: These are annual or perennial herbs with a rush-like appearance and milky juice. The flowering head is comprised of all ray flowers that are pink or purple with 4-8 involucral bracts. In general, the species are found in found in dry, open places at lower elevations. Two species, *L. exigua* (= *Prenanthella e.*) and *L. spinosa* (= *Stephanomeria s.*) are found in southern California.

Ecology & Ethnobotany: Some species were reportedly drunk as a tea by Native American women to increase their milk flow. For example, the bluish colored tea of a related species, *L. juncea*, was also believed to give the mother an "inner power" which was then passed on to the infant. The sap of *L. juncea* can be chewed like gum. A tea made from the whole plant is said to cure diarrhea. Additionally, the leaves of *L. grandiflora* were boiled with meat or mush. An infusion of the plant was used to treat sore eyes (Willard 1992).

Tarweed (*Madia*)

Description: Typically, these are annuals with a tar scent of varying intensity. The leaves are narrow, usually opposite below and alternate above. The flower heads are comprised of inconspicuous yellow ray flowers. Tarweed can be found at moderate elevations in open, grassy, or vernally moist areas. There are three species in the southern California mountains.

Common Madia (*M. elegans*) This glandular, sticky, heavy scented annual is found on dry slopes from 3,000 to 8,000 feet. Flowers from June to August.

Threadstem Tarweed (*M. exigua*) This slender annual is found at low to moderate elevations. Flowers from May to July.

Mountain Tarweed (*M. glomerata*) This strongly ill-scented species occurs in forest openings in the San Bernardino mountains and northward.

Slender Tarweed (*M. gracilis*) This coarsely hairy annual is quite common and grows on wooded hillsides. Blooms from April to August.

Ecology & Ethnobotany: The seeds of *M. glomerata* may be eaten raw, cooked or dried and ground into meal. The scalded seeds also yield a nutritious oil. All Tarweeds were used medicinally by old Spanish settlers. An oil of excellent quality was made from their seeds in this country before olives were readily available.

Strike (1994) indicates that Tarweed seeds were collected and stored until needed. Seeds were often used in making pinole by many California Natives. The Miwok

Indians of California pulverized Tarweed seeds and ate them dry. When Tarweed seeds had matured but the plants were still green, Hupa Indians burned the areas were the plants grew. The seeds were then gathered from the scorched plants and because they needed no further parching, they were crushed into flour. The roots of some species were also eaten.

Woolly Malacothrix (*Malacothrix floccifera*)

Description: This almost leafless annual has branched stems that are 4 to 16 inches tall. The stems contain a milky juice. Leaves are mostly basal, oblong, 3/4 to 3 inches long, pinnatifid, and toothed. The flowering heads contain many flowers that are all ray flowers, with white or pale-yellow ligules. These may be tinged with pink. The ligules are 1/8 to 3/8-inch long, and end in a 5 toothed apex. The pappus is of 1 to 8 bristles. Woolly Malacothrix grows on sandy and rocky places below 5,000 feet from Ventura County north. Flowers from April to June.

Ecology & Ethnobotany: Several Native American tribes ate the seeds of a related species (*M. californica*)

Pineapple Weed (*Matricaria matricarioides*)

Description: This is an erect annual herb that has a branched habit. Leaves are alternate and pinnately lobed or divided. The small, terminally arranged flower heads are composed of disk or ray flowers. This is a common species in waste places at the lower elevations. *Matricaria*, in general, are introduced plants that are

circumboreal in distribution. The scientific name is from the Latin mater or matrix, referring to the plants reputed medicinal value.

Ecology & Ethnobotany: A delicious tea can be made from the dried flowers and the leaves are edible, but bitter. The medicinal uses of Pineapple Weed are identical to that of Chamomile (*Anthemis*). It can be used as a tea as a carminative, antispasmodic, and mild sedative.

Nodding Silverpuffs (*Microseris nutans*)

Description: This annual grows 4 to 24 inches tall and has mostly basal leaves and a leafless stem. Leaves are 1½ to 10 inches long, linear to narrow elliptic in shape, and entire or with a few teeth or pinnatifid. The flowering heads are terminal on the stems. Involucres are 5/8 to 1½ inches long. The heads are comprised of all ray flowers, and the ligules are yellow. Achenes are not beaked and the pappus is of 5, silvery, papery scales that are tipped with an awn or bristle. Microseris grows in grassy places or in open woods below 6,000 feet. Flowers from April to June.

Ecology & Ethnobotany: The slender roots of Nodding Silverpuffs are apparently edible raw.

California Coneflower (*Rudbeckia californica*)

Description: This unbranched perennial grows up to 6 feet tall and has stems that are short hairy above. The leaves are alternate, lanceolate to elliptic in shape, and rough hairy. The lower leaves are irregularly toothed, while the upper leaves are entire and sessile. The flowering heads are solitary and terminal, large and showy. Both ray and disk flowers are present. The disk flowers are  conical, greenish yellow, and elevated above ray flowers. The ray flowers are yellow and there are 8 to 21 per head. California Coneflower is found in Kern County and northward from 5,500 to 7,800 feet. Flowers from July to August.

Ecology & Ethnobotany: Several species are suspected of poisoning livestock when eaten in quantity. A root tea from *R. hirta* (Black-eyed Susan) was used for worms and colds, and as an external wash for sores, snakebites, and swelling. The root juice can also be used for earaches. A root tea from *R. lacinata* (Cutleaf Coneflower) was drunk for indigestion, a poultice made from the flowers was applied to burns, and the cooked spring greens were eaten for "good health".

Groundsel (*Senecio*)

Description: They are annual, biennial, or perennial herbs with alternate or basal leaves. The flower heads are yellow and the pappus is made up of hair-like bristles. Groundsels can be found in various habitats and elevations. *Senecio* is one of the largest genera of plants with nearly 2,000-3,000 species distributed worldwide. Approximately 100 species are found in the Western United States.

Groundsel (*S. astephanus*) This perennial from an erect rhizome occurs on steep rocky slopes up to 5,000 feet in the Transverse Range.

San Bernardino Ragwort (*S. bernardinus*) This perennial is rare and occurs in pine forests of the eastern San Bernardino mountains from 4,500 to 7,000 feet.

Dwarf Mountain Butterweed (*S. fremontii*) This is a dwarf perennial plant grows in rocky areas from 8,500 to 12,000 feet in the San Bernardino Mountains. Flowers from July to August.

Tehachapi Ragwort (*S. ionophyllus*) This perennial herb is uncommon in dry, rocky coniferous forests and granite crevices in the San Gabriel and San Bernardino mountains between 5,000 to 8,000 feet.

Ragwort (*S. triangularis*) This is a glabrous perennial is common plant in wet meadows and along stream banks from 4,000 to 11,500 feet in the southern California mountains and northward. Flowers from July to September.

Ecology & Ethnobotany: Many species contain highly toxic alkaloids and should therefore be avoided. A related species, *S. douglasii* (Douglas'Groundsel), was used medicinally by southwestern Native Americans as a laxative, although misuse could result in death. Strike

(1994) indicates that young *Senecio* leaves were eaten by Maidu Indians in California as cooking herbs. Additionally, the seeds may have been eaten by the Chumash Indians (California). *Senecio* leaves were apparently used to line earth ovens.

Milk Thistle (*Silybum marianum*)

Description: The name milk thistle refers to the white streaks along the leaf veins. In Germany, where the plant is often depicted as a religious symbol associated with the Virgin Mary, legends describes the white mottling to a drop of the Virgin Mary's milk. The species name (*marianum*) honors the symbolic association of the plant with the Virgin Mary.

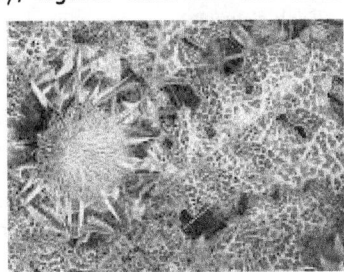

Ecology & Ethnobotany: This naturalized species from the Mediterranean was brought to America by early settlers, probably as food. The black shiny seeds are crowned with feathery tufts and have been roasted as a coffee substitute.

Medicinally, the plant has gained prominence based on research conducted in the past thirty years. However, it has been considered to be a "liver-protecting" herb since the 1st century.

Goldenrod (*Solidago*)

   Description: The various species of Goldenrod
are perennial herbs with fibrous roots. The leaves are
alternate, simple, and either tooth or entire. The heads
are made up of yellow ray flowers. Goldenrods may be
found in dry to moist habitats from the foothills to
timberline, often in dense patches. *Solidago* means "to
make whole."

"Quick Key" to the Goldenrods

1. Disk flowers 11-21 per head; lowermost leaves entire
and petioled; upper leaves linear-lanceolate - *S. confinis*
1. Disk flowers 4-11 per head; lowermost leaves finely
toothed, narrowed at the base - *S. californica*

   California Goldenrod (*S. californica*) This is a
perennial plant that is common in dry or moist fields
and in clearings or openings from 200 to 7,500 feet.
Blooms July to October.
   Southern Goldenrod (*S. confinis*) This stout
perennial grows in dry or moist banks up to 8,200
feet from Ventura to San Diego counties. Flowers
from July to October.

   Ecology & Ethnobotany: Young leaves can be
prepared as potherbs or added to soups. Depending on
habitat, age, and personal preference, their palatability
is quite variable. The dried leaves and dried, fully
expanded flowers can be used to make a tea. The seeds
can be used to thicken stews. Large amounts of the raw
herbage should be avoided as it may be toxic.

Medicinally, *Solidago* was employed for checking internal and external bleeding. An antiseptic lotion may be made by boiling the stems and leaves, or by using dry, powdered leaves. The powdered dry leaves were also sprinkled on cuts as a styptic. For insect bites and minor scrapes, apply fresh, crushed or chewed leaves. A tea wash is said to be good for rheumatism, neuralgia, and headaches.

The fluffy down from the flower heads are a good additive for tinder bundles. All Goldenrods contain small quantities of natural rubber, and were once cultivated as a domestic source.

Sow Thistle (*Sonchus*)

Description: The three species of Sow Thistle that may be encountered include *S. arvensis* (Field Sow Thistle), *S. asper* (Spiny Sow Thistle), and *S. oleraceus* (Common Sow Thistle). Introduced from Europe, they are weedy perennials and annuals with alternate leaves that are entire to pinnately divided. The leaf bases have eared-shaped lobes and the margins are prickly. The flower heads are composed of entirely yellow ray flowers. The pappus is bristly. The common name is said to be derived from the observation that pigs eagerly consumed the plants. In general, they occur at the lower elevations in gardens and waste places.

Ecology & Ethnobotany: The young plants of all three species can be prepared as a potherb. As they get older, they become increasingly bitter. We found that boiling them in at least two changes of water makes them a little more palatable. Since the plants have an abundance of soluble vitamins and minerals, use only a minimum amount of water and boil briefly. The milky gum obtained from *S. oleraceus* was once used in treating opium addiction, and Native Americans used a tea made from the leaves of *S. arvensis* to calm nerves (Foster and Duke 1990). In Europe, a poultice from the leaves was used as an anti-inflammatory.

Wire Lettuce, Skeleton Weed (*Stephanomeria*)

Description: These are more or less branched annual or perennial herbs with milky juice. The leaves are small and often scale-like. The flowers are pink and composed of ray flowers. The four species are found in dry, open places at low and mid elevations.

Chicory-leaved Stephanomeria (*S. cichoriacea*) This perennial plant occurs grows on rocky slopes and canyons in the San Bernardino Mountains and the Santa Ana Mountains and north below 6,000 feet. Blooms from August to October.

Small Stephanomeria (*S. exigua*) This annual grows in low, dry places below 8,500 feet. Flowers from May to September.

Wire Lettuce (*S. pauciflora*) This rounded bush or shrub grows in washed and in open areas below 6,000 feet. Flowers from May to September.

Tall Stephanomeria (*S. virgata*) This erect, stiff annual is found in dry, disturbed areas below 6,000 feet. Flowers from July to October.

Ecology & Ethnobotany: *Stephanomeria virgata* exudes a milky sap and was used as an eye medication by the Kawaiisu Indians. The sap of *S. pauciflora* was used as a chewing gum.

Dandelion (*Taraxacum officinale*)

Description: Dandelions need very little introduction. All species are tap-rooted perennials with milky juice and leaves that form a dense, basal rosette. The solitary flower head is composed of bright yellow ray flowers. They are found in a variety of habitats up to the alpine zone. *Taraxos* means "disorder", and *akos* means "remedy." The plant was an ancient remedy for ailments ranging from spring doldrums to mononucleosis. The latex-like sap was a folk medicine for warts. Red-seeded Dandelion (*T. laevigatum*), formerly considered a separate species is now referred to as *T. officinale*.

      Ecology & Ethnobotany: Common Dandelion was introduced into North America by European settlers as a food crop and a medicinal cure-all. Every part of Common Dandelion is edible. The young leaves may be eaten raw or cooked like spinach. The older leaves are also edible, but we find it is better to boil the older leaves in 1 or 2 changes of water to eliminate the bitterness that comes with age. The plants are high in Vitamins A and C, a good source of B complex, and iron, calcium, phosphorous, and potassium. The roots can also be eaten raw, or boiled as a vegetable, baked as potatoes, or added to soups and stews. The roasted root can be used as a substitute for coffee, but it lacks the caffeine buzz. The flower buds can be pickled and added to meals such as omelets. In general, Common Dandelion is good for blood circulation.

The less common native species of *Taraxacum* may also be edible, but no authenticated information has been located. Kirk (1975) also believes that the other native species have similar qualities. During World War II, a species of Dandelion was cultivated by Russians as a commercial source of rubber. In spite of its many uses, many Americans consider the Dandelion a pest and spend much time and money trying to eradicate it.

Note: The California Dandelion (*T. californicum*) is listed as a threatened species by the U.S. Fish and Wildlife Service and is found in the San Bernardino Mountains. Extra care should be taken to differentiate the common species from this rare one.

Horsebrush (*Tetradymia canescens*)

Description: This is a low, rigid shrub growing up to 12 inches tall. The stems are covered with a dense wool, and the leaves are alternate, narrow, entire, linear to linear lanceolate, and covered with white wool. The flowering heads have only disk flowers, and the involucres are white woolly. There are 4 flowers per head with yellow corollas. The pappus is yellowish with capillary bristles. Horsebrush grows on dry slopes from 4,000 to

10,000 feet in the San Bernardino Mountains and northward. Blooms from July to August.

Ecology & Ethnobotany: An infusion from the root of this species was used to soothe sore eyes. The bark of this and other species provided chewing gum, and the spines used as needles for tattooing.

Salsify (*Tragopogon porrifolius*)

Description: This is an introduced, tap-rooted, biennial herb with milky juice. The leaves are alternate, entire, sessile and clasping at the base and taper to a long point. The flower heads are solitary and composed of pale yellow or purple ray flowers. The heads open early in the day, close about noon and remain closed on cloudy, rainy days. They are found in many habitats at lower elevations. The genus name is Greek, for "goat's beard", probably referring to the thin, tapering, tufted, grass-like leaves.

Ecology & Ethnobotany: The fleshy roots of Salsify can be eaten raw or after cooking. The flavor resembles that of an oyster, an acquired taste! Other species, *T. dubius* (Yellow Salsify) and *T. pratensis* (Meadow Salsify), are also edible, but are somewhat smaller, more fibrous, and tough. Salsify root has been cultivated for over 2,000 years in the Mediterranean. The young leaves and stems of all species can be eaten after boiling until tender. The coagulated sap can be used as

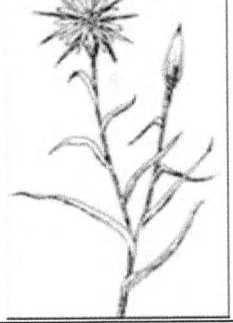

chewing gum and as a remedy for indigestion.

Wyethia (*Wyethia ovata*)

Description: This rough to the touch erect perennial has short stems, grows up to 12 inches tall, has large, ovate leaves that hide the heads of flowers. The leaves are alternate, broadly ovate to roundish. The flowering heads are hidden among the leaves. Both ray and disk flowers are present, but the ray ligules are short

and are not as long as the outer involucral bracts. There are about 5 to 8 yellow ray flowers per head. The disk flowers number 12 to 20, and the pappus consist of a crown of unequal scales. Wyethia grows in grassy, open, or wooded areas between 1,200 to 6,000 feet. Flowers from May to August. All species have leaves on the stems distinguishing them from *Balsamorhiza*, which only has leaves at the base.

Ecology & Ethnobotany: The seeds are edible and somewhat resemble sunflower seeds in taste. The roots of a related species, *W. helinioides* (Whitehead Wyethia), can be eaten after they have been cooked for a day or two, in a steam pit. Regardless of how long the roots are cooked, the smell is almost intolerable to enjoy eating. A decoction of leaves was used as a bath, producing profuse sweating. The leaves are considered to be poisonous and should not be taken internally. The Klamath Indians used the mashed root as a poultice for swellings.

Cocklebur (*Xanthium strumarium*)

Description: Two species, *X. spinosum* (Spiny Cocklebur) and *X. strumarium* (Rough Cocklebur) occur in southern California. These are coarse annual weeds of uncertain origin that have a cosmopolitan distribution. The stems of the plants are simple, and the leaves are alternate. The flower heads are solitary or clustered in the leaf axils. The bur (seed) has conspicuous, slender hooked prickles. Cockleburs can be found in the lower elevations.

Ecology & Ethnobotany: The uses of Cocklebur are primarily medicinal. They have been used by many aboriginal people throughout North and South America, and as a herbal medicine in China. The seeds were ground, mixed with corn meal, made into cakes or balls, and steamed by the poorer class of Zuni Pueblo.

Historically, the roots of *X. strumarium* were used for scrofulous tumors (related to the lymph glands in the neck). Both species were also used for rabies, fevers, and malaria. They possess diuretic, fever reducing, and sedative properties. Native Americans used a leaf tea for kidney disease, rheumatism,

tuberculosis, diarrhea, and as a blood tonic. The seeds have germicidal qualities, and were ground and applied to wounds. The seeds also contain an oil that can be used as lamp fuel.

BARBERRY FAMILY (Berberidaceae)

There are nine genera and 590 species in this family, distributed throughout the northern hemisphere and South America. Two genera are native to the United States. This is a diverse family of perennial herbs and shrubs. The flowers have six or more stamens that split open by two hinged valves to splatter pollen over insects as they crawl by. Several species are cultivated as ornamentals.

Barberry, Oregon-grape (*Berberis*)

Description: The genus *Berberis* has about 600 species; 6 grow in California. In California these are rhizomatous, viney, or upright shrubs with pinnately compound, evergreen leaves. The leaflets have spiny margins, and the yellow flowers are in 3 whorls that are interpreted as bracts, sepals, and petals. The fruits are blue to purple in color and have a waxy covering.

In some books, Barberries are those plants with simple leaves and spiny stems, whereas Oregon-grapes have compound leaves and spiny blades. As such, they have been treated as separate genera *Berberis* and *Mahonia*, respectively.

"Quick Key" to the Oregon-grape

1. Leaf petioles over 1 inch long and leaflets are not crowded - **B. aquifolium**
1. Leaf petioles less than 1 inch long and leaflets do not overlap each other nor are they crowded - **B. pinnata**

Oregon-grape (B. aquifolium) This is a creeping or erect shrub that may grow up to 10 feet tall. Occurs below 7,000 feet in forests. Oregon-grape take many forms that grade into each other. These forms have been treated differently by botanists with as many as 6 different species being recognized. Here we shall follow Stuart and Sawyer (2001) who recognize 1 species.

California Barberry (B. pinnata) This is an erect or vine-like shrub up to 6 feet tall. Occurs usually below 5,000 feet in the southern California mountains.

Ecology & Ethnobotany: The blue berries are edible raw or can be dried for winter use or added to soups to improve flavor. We like them best when picked right off the plant. They also make good jellies.

Medicinally, the bark may be boiled and the infusion used to wash sores on the skin and in the mouth. The plants contain berberine, a bitter alkaloid that gives roots their distinctive yellow color and usefulness as a digestive tonic. Berberine stimulates the involuntary muscles and possesses anti-pyretic, laxative, and anti-bacterial qualities. A tea from the berries of *B. vulgaris* (Common Barberry) can be drunk to stimulate an appetite.

The liquid obtained from the root by chewing was placed on injuries and wounds.

A yellow dye can be obtained by boiling bark and roots. The whitish film on the berries is a yeast that can be used in making a primitive sourdough starter.

Caution: The roots should be considered toxic, and the spines on the leaves may inject fungal spores into the skin.

Oil of Wintergreen

Oil of wintergreen (methyl salicylate) is a folk remedy for body aches and pains, and is known for its astringent, diuretic, and stimulant properties. Caution: Oil of wintergreen, used externally for pain (i.e., muscular, joint, arthritic and rheumatic), may cause irritation to the skin. As a precaution, test by using a few drops of wintergreen oil in a carrier oil or liniment.

BIRCH FAMILY (Betulaceae)

Of the six genera and 150 species in the Birch Family, five genera are native to the United States. Members of this family are trees and shrubs with deciduous, simple, alternately arranged leaves that have toothed margins. The unisexual minute flowers are arranged in catkins. Both male and female flowers are borne on the same plant. The male catkins are soft and pendant. After releasing abundant pale-yellow pollen in early spring, they drop off. The hard, female catkin is either erect or pendant, and appears cone-like. Most members of this family grow in moist soil, particularly along streams. Economic products of this family include lumber, edible seeds, and oil of wintergreen.

White Alder (*Alnus rhombifolia*)

Description: This is a tree growing 30 to 100 feet tall. It has grayish-white bark and the trunk is up to 24 inches in diameter. The toothed and finely pubescent leaves are alternate, ovate, and dark green and glabrous above, lighter green below. Male flowers occur in catkins and the fruit (female) is a woody cone - resembling a cone of a conifer. White Alder can be found in moist areas along streams below 5,000 feet. Flowers from January to April.

Ecology & Ethnobotany: The edible catkins are high in protein, but generally don't taste very good. The catkins are more tolerable if they are nibbled raw, added to soups, or dried and powdered and used as a spice. The inner bark is palatable only for a short time in the spring when it is less bitter. A patch of bark is removed from the tree and the tissue scraped off and eaten fresh or dried in cakes.

The bitter leaves and inner bark act on the mucous membranes of the mouth and stomach to stimulate digestion. A tea made from the leaves was used as a wash or a soothing remedy for Poison Ivy, insect bites, and other skin irritations. Used fresh, the inner bark is emetic, taken to induce vomiting if poisonous substances are ingested. A decoction from Alder bark was used to treat colds and stomach trouble. The decoction was also used to reduce pain from burns and scalds. Chewing alder bark is said to turn one's saliva red, which was used to dye basketry material.

Alder is valued for its hardwood and is useful for open fires as it does not readily spark. It is used widely by aboriginal peoples for woodworking, including dishes, spoons, and platters. The wood is also used for making fire drill sets. The astringent bark and woody cones are used for tanning leather. A black-brown dye from the bark was used for coloring fishing nets to make them more invisible. Since Alder usually grows in the vicinity of free-flowing water, it is considered a botanical

indicator of water. The roots have small nitrogen nodules that improve the soil for other plants. Alders are good for controlling erosion and floods, and for stabilizing streambanks.

Coal Burning

Have you ever tried to carve a depression in wood to make a spoon, cup, or bowl with only a knife? It can be frustrating at times To simplify the task, try using hot coals. Place a piece of coal where you want the wood to be hollowed-out, then blow on the embers with a steady, thin stream of air to keep the coal glowing. A straw is often helpful in directing the air stream. After the coals have burned down a bit, scrape out the charcoal with a knife. Repeat the process with fresh coals until the depression is formed.

BORAGE FAMILY (Boraginaceae)

The Borage Family has approximately 100 genera and 2,000 species, with 22 genera native to the United States. Members of the Boraginaceae can be identified by their alternate leaves, round stems, coiled racemes, and 5-parted radially symmetrical flowers. The corolla has a narrow tube that is abruptly flared at the top. The fruit, comprised of 4 nutlets, is helpful to correctly identify species in this family. The name borage comes from a Middle Latin source, *burra*, meaning rough hair or short wool as many of the plants in the family are covered with stiff hairs.

Fiddleneck (*Amsinckia menziesii*)

Description: Fiddlenecks are coarse annual herbs with stiff hairs. The flowers are in a scorpion tail-like spike. This species is usually found below 6,000 feet

in dry areas. Named for W. Amsink, an early 19th century patron of the botanic garden in Hamburg, Germany.

Ecology & Ethnobotany: The seeds of this species were pounded into flour, then made into cakes and eaten without cooking. Coastal California tribes also ate the young fiddleneck leaves. A related species, *A. douglasiana*, was apparently used medicinally (unspecified) (Strike 1994).

Catseye, Cryptantha (*Cryptantha*)

Description: The many species of Cryptantha are annual or perennial herbs that are rough to the touch. They have linear or spatulate leaves and the inflorescence is scorpion tail-like with small, white or yellow flowers. Popcorn Flowers are usually found in dry, open areas at various elevations. Four species of Cryptantha can be found in the southern California mountains.

"Quick Key" to the Cryptantha

1. Plant low and spreading; flowers in axils of bracts - *C. circumscissa*
1. Plant erect; flowers not leafy bracted - **2**

2. Corolla $\frac{1}{4}$-inch across; nutlets smooth and shining - *C. mohavensis*
2. Corolla less than $\frac{1}{4}$-inch; nutlets rough - **3**

3. Corolla 1/8 to 1/4 inch across; nutlets all alike without wings - *C. muricata*
3. Corolla 1/16 inch across; 3 nutlets with winged margins, one without - *C. pterocarya*

Cryptantha (*C. circumscissa*) This is a low, much branched annual grows in sandy or gravelly places below 9,000 feet, mostly in desert mountains. Flowers from April to August.

Cryptantha (*C. mohavensis*) This erect annual grows in dry places from 2,500 to 9,000 feet in the west Mojave Desert, to the Tehachapi Mountains and northward. Flowers from May to July.

Prickly Cryptantha (*C. muricata*) This is an erect, branched annual grows on gravelly or rocky slopes below 8,000 feet from Orange County northward.

Wing Nut Cryptantha (*C. pterocarya*) This erect annual grows in sandy or gravelly places below 6,000 feet on the desert side of the mountains. Flowers from March to June.

Ecology & Ethnobotany: The seeds of some *Cryptantha* may have been eaten by the Chumash Indians in California.

Stickseed (*Lappula redowski*)

Description: This is a taprooted annual herb with linear, alternate leaves. The plants are densely hairy throughout. The flowers are blue (rarely white) and the fruits have prickles about the edges. Stickseeds are usually found at the lower elevations in dry, disturbed habitats up to 8,000 feet in the San Bernardino mountains and northward. *Lappula* is Latin, diminutive of lappa, meaning bur.

Ecology & Ethnobotany: Medicinally, a poultice made from Stickseed was applied to sores caused by biting insects, whereas as cold infusion was used as a lotion for sores and swellings. Additionally, the roots of European Stickseed placed on a hot stone and allowed to smoke was used as an inhalant, and a snuff was made from the root for headaches

Popcorn Flower (*Plagiobothrys*)

Description: The species are annuals with alternate or opposite leaves. The white, salver-form flowers are in scorpioid racemes. The various species occur in moist soil.

"Quick Key" to the Popcorn Flowers

1. Leaves ¾-2 inches long; flowering spike without bracts; corolla 1/8 inch across; low elevations - *P. arizonicus*
1. Leaves ¼-¾ inch long; flowers in leafy bracted spike; corolla less than 1/8 inch across; high elevations - *P. torreyi*

Popcorn Flower (*P. arizonicus*) This is an erect annual grows on dry slopes at lower elevations, mostly in the desert areas. Flowers from March to May.

Popcorn Flower (*P. torreyi*) This is an erect annual that has a purple dye in the stems and roots and grows in meadows from about 4,000 up to 11,000 feet. It occurs mostly from the San Bernardino Mountains northward. Flowers from June to July.

Ecology & Ethnobotany: A rouge was obtained from the stem base and roots of some species of *Plagiobothrys*. Many species have purple dye in the stems and roots. When pressed and dried in a folded sheet of clean paper, a mirror image pattern often results. The shoots and flowers of related species in California (*P. fulvus*) were eaten, and leaves eaten as greens.

MUSTARD FAMILY (Brassicaceae)

There are approximately 375 genera and 2,000 species in this family worldwide. Mustard flowers are easy to recognize with 4 petals in the form of a cross, hence the name "cruciform." There are usually six stamens, four of which are longer than the remaining two. The fruit is a pod, either long and thin (silique) or short and wide (silicle), with a partition down the middle dividing the seeds into two individual chambers.

Mustards are a friendly family, having a characteristic peppery taste. In general, the genera are safe to experiment with. Mustards are associated with the Roman practice of soaking seeds in newly fermented grape juice ("must"), drunk as a stimulant to prepare armies for battle. Cauliflower, turnip, radish, cabbage, rutabaga, and watercress are among the economically important plants in this family. Despite the high nutritional value of mustards and tenacity of plants, they are largely neglected. In America, Mustards are often regarded as pests, but in Europe and Asia, various species are widely cultivated.

Rock Cress (*Arabis*)

Description: These are biennial or perennial herbs with stellate hairs. Flowers are in racemes, usually white to purple in color. The fruits are linear siliques, usually flattened parallel to the partition. The four species of *Arabis* in southern California are found in a variety of habitats at various elevations.

Brewer's Rock Cress (*A. breweri*) This perennial herb grows in rocky places, crevices, talus slopes in the San Bernardino mountains (e.g., Dollar Lake) from 5,000 to 9,600 feet.

Rock Cress (*A. dispar*) This perennial occurs in loose gravelly slopes and compact talus between

4,000 to 7,500 feet in the San Bernardino mountains.

Rock Cress (*A. holboellii*) This Rockcress grows in dry, stony places in the San Bernardino mountains between 6,000 to 11,000 feet. Flowers from May to July.

Rock Cress (*A. parishii*) This species grows on dry, stony slopes from 6,500 to 9,800 feet in the San Bernardino mountains. Flowers from April to May.

Rock Cress (*A. platysperma*) This species grows on dry, stony slopes from 5,500 to 11,200 feet in the San Gabriel and San Jacinto mountains. Flowers from June to July.

Ecology & Ethnobotany: The crushed plant of *A. puberula* (Silver Rockcress) serves as a liniment or mustard plaster. Some species of Rock Cress were eaten by California tribes, and an infusion was used to cure colds.

Wintercress (*Barbarea orthoceras*)

Description: This herb has angled stems and pinnatifid leaves. The flowers are yellow, and the pods are

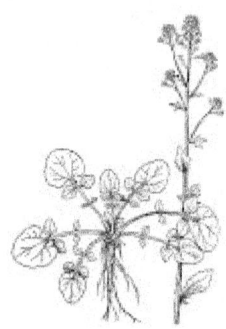

linear and 4-angled. The genus is named for Saint Barbara. The Latin name is derived from the fact that the young leaves can be eaten on St. Barbara's Day in early December. Wintercress grows in moist places and along stream banks from 2,500 to 11,000 feet. Flowers from May to September.

Ecology & Ethnobotany: The young stems and leaves of Wintercress and *B. vulgaris* (Garden Yellowrocket) can be eaten raw in salads or prepared as a potherb. We like to boil the plants in at least two changes of water.

Mustard (*Brassica*)

Description: *Brassicas* are large annuals with showy yellow flowers. The pods are round or 4-sided in cross section, with a conspicuous beak. In southern California, the various species can be found in waste places and fields at lower elevations. *Brassica* is the Latin name for cabbage.

Ecology & Ethnobotany: All species of Brassica have leaves that are edible as greens - an excellent source of Vitamin A, B, and C, calcium and potassium. The older leaves should be boiled in at least one change of water. The seeds contain thiocyanate and may cause goiter if consumed in large amounts. The seed can be ground or crushed to a flour and applied as a mustard plaster. The plaster is a long-used remedy for aches and pains. Mustard oil is also a caustic irritant and can discolor and blister the skin if left on too long. The flower buds are rich in protein. The table mustard comes from *B. nigra* (Black Mustard). In China, mustard oil from *B. rapa* (Rape Mustard) was used for illumination until the introduction of kerosene.

Shepherd's Purse (*Capsella bursa-pastoris*)

Description: This is a pubescent annual with leaves mostly in a basal rosette. The petals are white and the pods obcordate (heart-shaped), strongly flattened contrary to the narrow septum. *Capsella* means "little box," referring to the fruit, as does *bursa-pastoris*. Collectively, the name means "purse of the shepherd." This is a common weed on dry or disturbed soil. Blooms most of the year.

Ecology & Ethnobotany: Shepherd's Purse has been used as food for thousands of years. The seeds have been found in the stomach of Tollund man (approximately 500 BC - 400 AD) and during excavations of the Catal Huyuk site, approximately 5950 BC. The seeds of

Shepherd's Purse may be parched and eaten or ground into flour. The whole pod, with the seeds beaten out, can be added to salads or soups or dried for winter use. The young leaves can be prepared as a potherb and are a good source of Vitamin C. With age, they develop a peppery taste. The entire herb (leaves, stems, green seed pods) can be chopped and added to soups. The roots may be ground or chopped and used as a ginger substitute. The seeds are known to cause blistering of the skin.

The plant is extremely high in Vitamin K, the blood clotting vitamin. Mash or chew the leaves and hold them on a cut. The juice of the plant on a ball of cotton was used to stop a nosebleed. Shepherd's Purse also contains significant amounts of calcium, potassium, sulfur, and ascorbic acid. Used as a decoction, Shepherd's purse has been used to treat hemorrhoids, diarrhea, and bloody urine. The decoction has a gentle detergent action, and is very cleansing to the skin.

Biological Control of Mosquitoes

The seeds of Shepherd's Purse have been used against mosquitoes. By sprinkling the seeds on water where mosquitoes breed, the mucilage on the seeds will kill the larvae. One pound of seeds is said to destroy ten million larvae. In water, the seeds release a mosquito attractant and gummy substance that binds the mouth of larva to the seed. In addition, the seed releases a substance that destroys larvae. Despite their pesky nature to humans, mosquitoes do have an important ecological role in nature, providing food for many fish, birds, and other insects.

Bittercress (*Cardamine breweri*)

Description: This plant has entire or pinnate leaves. The flowers are white or purple. The pods are elongate and flattened. The genus name comes from the Greek, because some Bittercress were thought to have heart strengthening qualities. Bittercress grows in the San Bernardino mountains in moist shady places and along streams from 4,000 to 10,000 feet. Flowers from May to July.

Ecology & Ethnobotany: The plants can be eaten raw in salads; however, we suggest cooking them in at least a change of water to improve the taste. Some plants in this genus were reputed to have medicinal qualities that were used in the treatment of heart ailments.

Caulanthus (*Caulanthus*)

Description: These are mostly annual or sometimes perennial herbs. Flowers occur in racemes and are white, purple, or yellow in color. There are approximately 18 species in the western North America. The genus name is from the Greek (*kaulos*), meaning stem, and (*anthos*), for flower - referring to cauliflower since some species can be used like it.

Caulanthus (C. amplexicaulis) Caulanthus grows on dry slopes from 5,000 to 8,500 feet in the San Bernardino and San Gabriel mountains, and Mt. Pinos. Flowers from May to July.

Wild Cabbage (C. coulteri) This species grows on dry slopes below 5,000 feet from Los Angeles county northward. Flowers from March to May.

Ecology & Ethnobotany: he leaves of C. coulteri can be eaten.

Tansy Mustard (*Descurainia*)

Description: Tansy Mustards are annual or biennial herbs with leaves that are 1-3 times pinnately divided. The foliage is covered with simple, branched, or short gland-tipped hairs. The flowers are cream-colored or light yellow and the pods are long, narrow, 3-sided to nearly round in cross section. These are weedy species occurring in disturbed soils at the lower elevations.

Tansy Mustard (D. incisa) This Tansy Mustard grows on dry slopes from 5,000 to 11,000 feet. Flowers May to August.

Tansy Mustard (D. pinnata) This species grows on dry, sandy areas below 8,000 feet. Flowers from March to June.

Ecology & Ethnobotany: All species of Tansy Mustard are edible as greens, but are bitter. The seeds can be parched, ground and prepared as mush. The seeds were parched by tossing in a basket with hot stones or live coals, then ground into a fine flour and made into mush. Because of its peppery taste, the mush was often mixed with the flour of other seeds to make it more palatable. Young leaves can be boiled or roasted between hot stones and eaten as green vegetables. The seeds were also used in poultices for wounds. However, one species, *D. pinnata* (Western Tansy Mustard) is reported to be poisonous in large quantities, causing blindness and paralysis of the tongue.

Whitlow Grass, Draba (*Draba corrugata*)

Description: This low, pubescent perennial has basal leaves in a tuft or cushiony rosette. The leaves are lanceolate and the flowers occur in a terminal raceme. The small flowers are white or yellow that fade to white with age. The pods are egg-shaped, elliptical, or club-shaped and sometimes twisted. Draba occurs on shaded slopes and rocky areas between 7,000 to 11,000 feet. Flowers from July to August.

Ecology & Ethnobotany: The species are mainly unpalatable. Whitlow grasses were formerly used for treating "whitlows," inflammations of the fingertip, particularly next to the nail.

Wallflower (*Erysimum capitatum*)

Description: Wallflowers are annual, biennial, or perennial herbs that are often tap-rooted. The herbage is covered with closely appressed forked hairs and yellow or orange flowers are showy. The linear pods are 4-sided in cross section with a small beak. Wallflower is common

on dry slopes below 8,000 feet. Flowers from March to July.

Ecology & Ethnobotany: Wallflowers were once used as a poultice. *Erysio* means to draw out, as in drawing out pain or causing blisters. Additionally, an infusion of dried, pulverized *E. capitatum* (Sanddune Wallflower), was rubbed on the head and face, to prevent sunburn, or to alleviate heat exposure. A pneumonia victim was cured by having his back massaged with chewed Wallflower root.

Peppergrass (*Lepidium*)

Description: The 13 species of Lepidium are annual or perennial plants that are widely distributed. One species is grown for salad. The genus name is Greek (*Lepidion*) and refers to a little scale with reference to the shape of the pods.

Peppergrass (*L. lasiocarpum*) This low, branched annual is common on grassy slopes and sandy areas below 5,000 feet. Flowers from February to May.

Virginia Pepperweed (*L. virginicum*) This is a widespread species found in waste places below 7,000 feet. Flowers from March to August.

Ecology & Ethnobotany: The young stems and leaves may be eaten raw or dried for future use. The plants contain Vitamins A and C, iron, and protein. The seed pods and seeds can be used as a flavoring. Fresh *L. virginicum* (Virginia Pepperweed) plants were bruised or a tea made from leaves was used for poison ivy rash and scurvy.

Wild Radish (*Raphanus sativus*)

Description: Wild Radish is a branched herb up to three feet tall. The flowers are white with rose or purple veins, sometimes yellowish. The fruit is a rounded pod up to two inches long. Wild Radish is a weed of waste places and fields at the lower elevations. The garden radish is a cultivated form of this species.

Ecology & Ethnobotany: The leaves, flowers, and pods of Wild Radish are used as a food, rather than the root. The root of Wild Radish is a bit too woody to be eaten like garden radishes. The flowers can be tossed in a salad or eaten alone as a snack. The fruits can be used in a salad, but must be collected before the seeds harden and the pods dry out. The taste of these pods resembles that of the garden radish.

Watercress (*Rorippa nasturtium-officinale*)

Description: This is an aquatic perennial that is slightly juicy or succulent. Leaves are pinnate compound into 3 to 11 ovate leaflets. The flowers are white or yellow, and the fruit is a curved pod. Watercress is common in quiet streams or on wet banks below 8,000 feet. Flowers from March to November.

Ecology & Ethnobotany: The herbage of Watercress is edible if the waters in which they grow are

not polluted. However, finding unpolluted water may be difficult. One suggestion would be to soak the fresh greens in a disinfectant, or treat the water with water purification tablets, or a tablespoon of bleach in a quart of water. Then rinse the greens well in potable water to remove the chemicals. The peppery-tasting plants were eaten raw or cooked as a potherb. A good source of vitamins, Watercress is listed as efficient in preventing scurvy. Medicinally, the plant was used for freckles, pimples, liver, and kidney troubles.

Note: Bittercress (*Cardamine brewerii*) looks similar to Watercress. To quickly differentiate the two species, look at the fruits. Bittercress fruits are linear and narrow; Watercress fruits are round or 4-angled in cross section.

Hedge Mustard (*Sisymbrium*)

Description: These are annual, biennial or perennial herbs. The small flowers are yellow or white, and the fruits are linear. The six species in California are introduced from Europe and are widespread throughout the United States. They are usually found in waste places and disturbed habitats at low elevations.

Ecology & Ethnobotany: The seeds of *S. officinale* (Hedge Mustard) can be parched and then ground into a flour. The plant also makes for a fine potherb. As with other mustards, it is best to cook the plants in a couple of changes of water.

Prince's Plume (*Stanleya pinnata*)

Description: This species has flowers that occur in elongated racemes. The fruits (siliques) are borne on a long stipe. The plants are usually found in sagebrush habitats at low elevations. The genus is named for Lord Edward Stanley, a British ornithologist who lived from 1775 to 1851.

Ecology & Ethnobotany: The tender leaves and stems of all four species can be prepared in much the same way as cabbage. They are bitter, but boiling in several changes of water remove some of the astringent properties. The seeds can be collected, parched, and then ground into a flour. They can be eaten as a mush or used in making breads.

Pennycress (*Thlaspi arvense*)

Description: The flowers are white; the flattened pods are egg-shaped to narrowly elliptical with a notch at the top. The non-native species occur in disturbed areas at the lower elevations, whereas the native species occur in open areas up to the timberline.

Ecology & Ethnobotany: The young shoots of can be eaten raw in salad, or cooked as a potherb. The plant has a mustard-like taste, with a hint of onion. Like other members of the mustard family, it's probably best to boil in at least two changes of water. Field Pennycress is high in Vitamin C and contains a relatively large amount of sulphur. The species may be toxic in large quantities.

Sand Lacepod (*Thysanocarpus curvipes*)

Description: This is a slender branched annual with stem leaves and basal leaves arranged in a rosette. The flowers are purplish, and the circular, flattened pods are surrounded by a flat nearly circular wing. The species occurs in open areas at low to mid elevations.

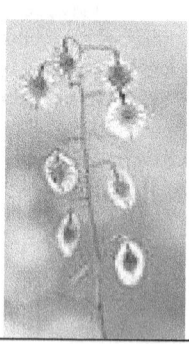

Ecology & Ethnobotany: Sand Lacepod seeds may be parched and eaten or ground into flour. A tea made from the plant is said to cure stomach-ache; a drink made from the leaf can be used to relieve colic.

WATERSHIELD FAMILY (Cabombaceae)

Two genera and eight species in the Watershield Family are distributed in temperate and tropical America, Africa, east Asia, and Australia. They are aquatic perennials typically found in freshwater. The genus is sometimes placed in the Nymphaeaceae (Waterlily Family), but differs in having simple pistils.

Watershield (*Brasenia schreberi*)

Description: Watershield is anchored to muddy substrates by slender rootstocks. All exposed portions of the plant are covered with a gelatinous sheath. The leaves are nearly round and arise near the tops of the stems. The flowers have purplish petals and sepals.

Ecology & Ethnobotany: The starchy rootstalk of Watershield can be peeled, boiled, and eaten, or dried and stored or ground into flour. The unexpanded young leaves and leaf stems can also be eaten in a salad, slime and all. The rootstalks were used to cure dysentery and stomach-ache.

CACTUS FAMILY (Cactaceae)

There are over 140 genera and 2,000 species within this family worldwide. Approximately 16 genera are native to the United States. Native to the Western Hemisphere, cacti have been spread all over the world, frequently carried by explorers and other travelers. Cacti are typically succulent spiny herbs of diverse form. One distinctive feature is the presence of areoles, which are round or elongated spots or openings that may be raised or pitted and usually arranged in rows or spirals over the surface of the plant. The spines grow from the areoles. The flowers have many sepals, petals, stamens, and an inferior ovary. Economic products from this family include ornamentals, edible fruits, nopalitas, and the hallucinogenic peyote. The spines of some cacti were once used as phonograph needles.

Prickly Pear (*Opuntia*)

Description: Prickly Pear need very little introduction. The various species in southern California are succulent herbs with fibrous roots and the stems that are flat or cylindric. The leaves when present are small, fleshy, and awl-shaped. The genus name may be from the Papago name (*opun*) for this food plant, or named for a spiny plant of Opus, Greece. Many *Opuntia* species have glochids - minute, nearly invisible barbed hairs that grow in clusters in areoles. They easily become embedded in skin or clothing, and, because of their light tan or yellowish color and barbed surface, are almost impossible to remove. Prickly Pear cacti occur in dry soils at the lower elevations.

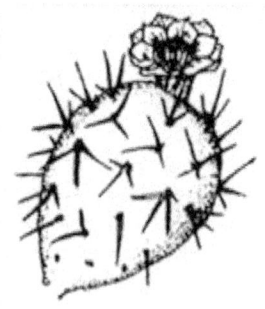

Ecology & Ethnobotany: Cacti have provided Native Americans with food, medicine, dyes, and a variety of other uses for thousands of years. They have also been attributed with saving many lives by supplying both food and water to people stranded in the desert. Since eating too many cactus fruits can cause constipation, they should be eaten in moderation. The cactus also contains oxalic acid, so watch your intake as you may develop a deficiency of calcium.

The pads, especially younger pads, make an excellent cooked vegetable. Harvest the young pads by grasping them with tongs and slicing them at the stem joints. Hold over a flame to singe off the spines and glochids, and scrape off the remaining ones with a knife. Rinse well. Slice into thin strips and boil for at least 10 minutes. Drain off water, rinse to remove the slippery gum and they're ready to eat. To use the older pads, slice away the more fibrous section, then cook accordingly.

After removing the spines, cactus fruits (also known as prickly pears) can be peeled, and the pulp eaten raw, or boiled and then fried or stewed. One solution for removing the spines is to burn them off; another is to split the fruit into two halves and eat the insides. The pulp can also be sun or fire dried for future use. High in protein and oil the nutritious seeds may be eaten or dried and ground into flour. Grind the seeds into flour or add them to soups.

Prickly Pear cacti pads have been used as a soothing poultice that can be applied to wounds and bruises. A tea made from the flowers was said to increase urine flow, and a tea from stems was used as a wash to ease headaches, eye troubles, and insomnia. The fruits are high in calcium, potassium, and Vitamin C. Additionally, the Chuvash Indians in California used large baskets baited with crushed Prickly Pear pads to catch sardines. Another tribe roasted *Opuntia* stems, then soaked them. The resulting extract was used to improve the plasticity and cohesion of clay when making pottery.

Archeologists in Texas have discovered purses made from prickly pear pads. According to Tull (1987), the dried pads were hollowed out to form a small container. A dye from the juice of the uncooked fruit can also be obtained. Bryan (1978) suggests letting wool soak for about a week in the fermenting juice. The color ranges from pink to magenta and appears to fade when exposed to sunlight.

WATER STARWORT FAMILY (Callitrichaceae)

Members of this family are small annual or perennial herbs with slender, usually lax stems. The leaves are simple, entire, and opposite or whorled. The minute unisexual flowers are borne in the axils of the leaves. There are no sepals or petals. The small, four-lobed fruit splits into four sections upon maturity. The plants are inconspicuous in standing water or drying mud. There is only one genus (*Callitriche*) and approximately 40 species in this family.

Water Starwort (*Callitriche verna*)

Description, Ecology, & Ethnobotany: This is the only genus of flowering plants in which aerial, floating, and subsurface pollination systems have all been reported. This species occurs in shallow water or on mud up to 11,000 feet elevation in the San Bernardino mountains. Flowers in May to August. The genus name is from the Greek, *kallos*, which means beautiful, and *trichos*, which means hair, referring to the slender stems. Strike (1994) indicates that the Maidu Indians used *Callitriche* to relieve urinary problems. The method is not reported, however.

BLUEBELL FAMILY (Campanulaceae)

Worldwide, there are over 70 genera and 2,000 species in this family. Twelve genera are native to the United States. These are annual or perennial herbs usually with milky juice. The flowers are typically 5-parted with the calyx divided into separate sepals, and the corolla is five lobed and bell-shaped. The family is of little economic importance, but some species are cultivated as ornamentals.

Scarlet Lobelia (*Lobelia cardinalis*)

Description: This is a simple, erect perennial growing 12 to 40 inches tall. Leaves are alternate, lanceolate, and 2 to 5 inches long with toothed margins. The lower leaves are petiolate and the upper leaves, sessile. Flowers occur in a terminal raceme that is 4 to 12 inches long. The flowers are irregular, bright red, 1 to 1½ inches long. The ovary is inferior and the anthers are united into a tube around the style. The fruit is a10 ribbed capsule that is 3/16-inch long. Lobelia grows in boggy places below 6,000 feet in Los Angeles, San Bernardino, and San Diego Counties.

Ecology & Ethnobotany: The various species of Lobelia have been used as drugs in many Native American Tribes. This species was used by the Cherokee as an analgesic where a compound was given for pain and a poultice of crushed leaves was used for headaches. An infusion of the leaves was given for rheumatism as well as for colds. As a panacea, the plant was used for ailment.

Caution: This is a toxic plant, especially when used as a home remedy.

Venus Looking-glass (*Triodanis*)

Description: These are annuals with fibrous roots. The flowers are solitary or several in the axils of the leaf-like bracts. The two species that may be encountered in the southern California mountains include *T. biflora* and *T. perfoliata*. Both occur in disturbed areas below 6,000 feet and are relatively uncommon.

Ecology & Ethnobotany: *T. perfoliata* was used by the Cherokee as a drug plant. For example, an infusion of the roots was taken and used as a bath for dyspepsia.

HEMP FAMILY (Cannabaceae)

The Hemp Family consists of annual herbs or climbing perennial herbs. The leaves are opposite, simple or compound. There are two genera - *Cannabis* and *Humulus* - with 3-5 species distributed in the north temperate zone, and they are widely cultivated. *Humulus* is native to the United States. The family is a source of hempen fiber, oils, edible seeds, hops, and tetrahydrocannabinols (THC), the psychoactive compound in *Cannabis*. Some references place the family in the Moraceae (Mulberry Family).

Indian Hemp, Marijuana (*Cannabis sativa*)

Description: This is an unbranched, coarse, aromatic annual. The lower stem leaves are opposite and the upper ones are alternate. The leaves are palmately compound with 3-9 leaflets that are lance-shaped to elliptic, coarsely toothed. The small, green male and female flowers (staminate and pistillate flowers, respectively) are found on separate plants. The fruit is an achene which is enclosed in the calyx and covered by a persistent bract.

Ecology & Ethnobotany: The plant is a native of Asia and was cultivated in Europe for fiber (hemp) and seeds (hemp butter, oil). The seeds can be used as food, parched and mixed into a batter and fried into cakes. Kephart (1964) indicates that the young shoots were used as a substitute for asparagus in Belgium, but because the plant contains poisonous alkaloids it should be avoided.

While the plant is now illegally cultivated in the United States and elsewhere for its euphoria-inducing properties, Marijuana has been legitimately used to treat glaucoma, and relieve nausea following chemotherapy. Although much maligned, Marijuana is potentially a very useful medicinal plant.

Warning: Cultivation of this species and use as a drug is still forbidden by federal law. It is usually grown by permit only under very rigid controls.

Common Hop (*Humulus lupulus*)

Description: The genus was formerly included in the Mulberry Family (Moraceae). Common Hops is a strongly twining, herbaceous vine with stems up to 15 feet long. The stems and leaves are rough to the touch. The leaves are opposite, serrate, 3-7 lobed, with heart-shaped bases. The underside of the leaves is glandular. The flowers are small, green, and unisexual. This is a widely cultivated plant from Europe and Asia.

Ecology & Ethnobotany: Hops are primarily grown for their fruits, used in brewing to give ale and beer a distinctive bitter taste. Additionally, the young shoots can be prepared as potherbs. They can be boiled in water for about 3-5 minutes, then boiled again in fresh water until tender.

A tea from the fruits was traditionally used as a sedative, antispasmodic, and diuretic, and used for insomnia, cramps, coughs, and fevers. Externally, the tea was used for bruises, boils, inflammations, and rheumatism. Recently, clinical studies have disproved the sedative qualities of Hops and found that it has no physiological activity on the nervous system; yet, anyone who drinks much of the tea tends to fall asleep or become groggy (Moore 1979).

Caution: The plant is known to cause dermatitis when handled.

CAPER FAMILY (Capparaceae)

Members of this family are shrubs, trees, or rarely herbs. They have simple or palmately compound leaves. Flowers have 4 sepals and 4 petals with the fruit a capsule or berry. There are about 46 genera and 800 species distributed in the tropical and subtropical areas of the world. Eight or nine genera are native to the United States. The family is of economic importance as a source of ornamentals and capers, a salad seasoning.

Bee Plant (*Cleome serrulata*)

Description: This is an erect, showy plant up to 40 inches tall with alternate leaves divided into three lance-shaped, entire leaflets. The reddish-purple to pink flowers are arranged in a dense, narrow, terminal inflorescence. The petals are separate while the sepals are united. The fruits are long-stalked, pendulous capsules, linear to lance-shaped in outline. Bee Plant is found in disturbed areas (i.e., roadsides, railroad right-of-way) at the lower elevations.

Ecology & Ethnobotany: An important food for many western Native Americans, Bee Plant was extensively used as a potherb. The young tender shoots and leaves, and flowers are preferred. The plant has an unpleasant odor, especially when older, and a pungent taste much like the mustards. We found it necessary to cook the plants in at least two changes of water to remove the bitter taste. The seeds can also be collected and ground into flour.

The Blackfeet Indians used the whole plant to make a medicinal tea to alleviate fever. Bee Plant may have been used by Native Americans to treat stomach-aches.

As a dye, plants are collected in quantity and boiled down for several hours until a thick, fluid residue is produced. The water is then drained off and the plants allowed to dry and harden into cakes. When black dye

or paint is needed, a piece of the cake is soaked in hot water.

HONEYSUCKLE FAMILY (Caprifoliaceae)

There are 15 genera and 400 species in the Honeysuckle Family. Of the 15 genera, 7 are native to the United States. They are woody plants with opposite leaves. The flowers are 5-merous, with the petals fused, and an inferior ovary. Many genera in this family are cultivated as ornamentals.

Honeysuckle (*Lonicera*)

Description: The two species here are shrubs or woody vines with entire and opposite leaves. They can be found in a variety of habitats from the foothills up to the alpine zone. The genus is named for Adam Lonitzer, a German naturalist who lived from 1528-1586.

"Quick Key" to the Honeysuckles

1. Uppermost leaves are fused into one - *L. interrupta*
1. Uppermost leaves distinct - *L. subspicata*

Chaparral Honeysuckle (*L. interrupta*) This twinning and trailing vine or shrub grows on dry slopes from 1,000 to 6,000 feet. Flowers from May to July.

Southern Honeysuckle (*L. subspicata*) This twinning and trailing vine or shrub is common on dry slopes below 5,000 feet. Flowers from April to July.

Ecology & Ethnobotany: The berries of Honeysuckle are seedy, but can be eaten raw, or dried for future use. The bark and twigs of a related species, *L. involucra*, were used for a variety of medicinal preparations, ranging from digestive tract problems to use as a contraceptive. Additionally, the juice from the stems was used as an antidote for bee stings.

The berries provide a black pigment. The long stems of Honeysuckle were used as basket foundation material by a number of Native American tribes. They also peeled and split the hairy stems as wrapping material for coiled baskets.

Elderberry (*Sambucus*)

Description: Elderberries are shrubs with pithy stems. The two species here have large, compound leaves with serrated leaflets. The white flowers are arranged in dense clusters. The fruits may be red or blue-black. Elderberries can be found in open areas, hillsides, and riparian habitats in the montane zone. The genus name comes from the Greek *sambuke*, an instrument made from the hollow stem.

Blue Elderberry (*S. mexicana*) This shrub grows 6 in open places up to 11,000 feet. Blooms from June to August.

Red-berried Elderberry (*S. racemosa*) This shrub grows in moist places between 6,000 and 11,000 feet in the San Bernardino mountains. It blooms from June to August.

Ecology & Ethnobotany: The blue or black elderberry berries of Blue Elderberry are edible raw, or they can be made into excellent jams, jellies, and wine. They can also be dried and stored for winter use. The seeds contain hydrocyanic acid, and if eaten in quantity can cause diarrhea and nausea. It is best to cook the berries or strain the seeds before use. The red-berried species contains much higher concentrations of these compounds and should be considered poisonous.

The blossoms can be added to pancakes to lighten batter and add flavor. The dried flowers were also ground and added to flours and baking mixes. Flower buds can be pickled or steamed as a potherb. Both the flowers and fruits contain a rich source of Vitamin C.

The fresh flowers can be used externally as a decoction for an antiseptic wash. Flower tea contains a natural estrogen and is often effective for relieving menstrual cramps. The leaves were used as poultices for sprains and skin irritations. The leaves and flowers were common ingredients in skin salves for piles, burns, and boils. Recent studies of Elderberry have confirmed that the berries possess antiviral properties that may be useful against influenza.

Elderberry stems can be cut and dried for use as musical instruments. After drying, holes can be bored into the branches to make flutes. During the drying process, the poisons are said to dissipate. The stems can also be used in making bows and arrow shafts in hunting small game. The odorous leaves can be used in water and sprayed on plants to repel aphids. The pith of the stem is used by watchmakers to absorb grease and oil. The leaves, with chrome as a mordant, yield a green hue. The berries, with alum and cream of tartar, yield a crimson dye.

CAUTION: The seeds, leaves, bark, and roots contain hydrocyanic acid and an alkaloid sambucine. They are toxic and cause acute emetic and laxative effects. Berries should be consumed when ripe and used for food after cooking and removal of seeds.

Snowberry (*Symphoricarpus rotundifolius*)

Description: This is a low, deciduous spreading shrub with branches up to 3 feet tall with gray-green leaves that are pubescent, simple, opposite, and ovate. The flowers are 2 to several, with tubular to bell-shaped pink corollas. The fruit is a white berry. Snowberry grows on dry, rocky slopes from 4,000 to 11,000 feet from San Diego County north to Santa Barbara County. Blooms from June to August.

Ecology & Ethnobotany: The white, tasteless berries are edible raw or cooked, and are said to be emetic and cathartic in large amounts. Saponins are found in the leaves and can be used as a natural

cleaning agent. A decoction of the pounded roots has been used for colds and stomachache. An infusion made from *S. rivularis* (= *S. albus* var. *laevigatus*) (Common Snowberry) was used to cure sores and skin lesions, and a root decoction was used to alleviate colds and stomach ailments.

PINK FAMILY (Caryophyllaceae)

There are approximately 80 genera and 2,000 species in this family found in the north temperate zone. About 20 genera are native to the United States. In general, they are annual or perennial herbs with opposite, simple leaves. The stems are often swollen at the joints. The flowers are 5-merous, and the calyx is tubular or has distinct sepals. Petals are often deeply notched, appearing like ten petals. Some species are cultivated as a source of ornamentals, and several genera are regarded as weedy.

Sandwort (*Arenaria*)

Description: The species in this genus are generally annual or perennial herbs with opposite leaves. The white flowers are borne in open to congested, flat-topped inflorescences. The species in southern California are found in various habitats from the foothills to above timberline in dry, rocky, and open areas, as well as moist open forests. The genus name is from the Latin, *arena*, referring to sand, the habitat of many species.

Sandwort (*A. lanuginosa*) This perennial with slender, spreading stems grows in moist places between 6,000 to 8,200 feet. Flowers from July to August.

Arenaria (*A. douglasii = Minuartia douglasii*) This delicate annual occurs in dry areas below 7,000 feet. Flowers from April to June.

Mojave Sandwort (*A. macradenia*) This is a

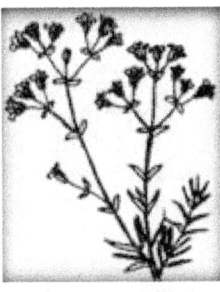

matted perennial herb that grows on dry, rocky slopes below 6,500 feet. Blooms from May to June.

Sandwort (*A. nuttallii = Minuartia nuttalli*) This species occurs on dry granitic areas between 6,500 to 11,000 feet in the San Gabriel and San Bernardino mountains. Flowers July to August.

Sandwort (*A. rubella = Minuartia rubella*) This species occurs in dry rocky places above 11,000 feet on San Gorgonio in the San Bernardino mountains. Flowers July to August.

Sandwort (*A. ursina*) This species is found on dry slopes between 6,000 to 7,000 feet in the San Bernardino mountains. Flowers June to July. *__This is an endangered species! Please avoid!__*

Ecology & Ethnobotany: The Sandworts are known for their medicinal uses by several Native American tribes. The roots of *A. aculeata* (Prickly Sandwort) was used as a decoction by the Shoshone as an eyewash, whereas a poultice of steeped leaves of *A. congesta* (Ballhead Sandwort) was applied to swellings.

Mouse-ear Chickweed (*Cerastium glomeratum*)

Description: This species of Chickweed is an annual with opposite, narrow, ovate leaves. The herbage is usually hairy and sticky. The flowers are white and the petals are deeply lobed at the tip. The fruit is a cylindrical capsule, often slightly curved at maturity. This is a common plant of waste places at lower elevations. The perennial species of *Cerastium* occur in dry to moist, open habitats up to the alpine zone. The genus name is from the Greek (*keras*), meaning horn, referring to the tapered capsule, which in some species is bent slightly like a cow's horn.

Ecology & Ethnobotany: *Cerastium* is frequently confused with *Stellaria media* (Chickweed), but to the general forager there is no danger. The tender leaves and stems of most *Cerastium* can be added to a salad, but we found they are better if boiled first and served as greens.

Bouncing-bet, Soapwort (*Saponaria officinalis*)

Description: There is an erect perennial herb with sessile or nearly sessile leaves. The flowers are showy, usually pale pink. Soapwort can be found along roadsides, disturbed areas, and waste places at the lower elevations. The plant has escaped from cultivation. The genus name is Latin for soap, since the juice of the plant lathers with water. Flowers June to September.

Ecology & Ethnobotany: The plant contains saponins and will irritate the digestive tract if eaten. The crushed green plant and roots can be used as a soap substitute.

Saponins and Soap

Many plants contain saponin which, once extracted, can be used as soap. Saponins can be found in Soapwort (Saponaria), Clematis (Clematis), Snowberry (Symphoricarpus), Elderberry (Sambucus), and other species. Cuts, wounds, and rashes need to be cleansed of bacteria to prevent infections, and when hunting, by eliminating the human scent we can avoid alerting game animals of our presence. Food and cooking gear also need to kept clean for health reasons. On the other hand, soaps can be detrimental to some living organisms living at the surface of water in ponds or streams. In water soaps break down the surface tension making life more difficult for organisms which are dependent on that surface tension.

Catchfly (Silene)

Description: There are many species of *Silene* in southern California. They are annual, biennial, or perennial herbs with opposite leaves. The sepals are united and often inflated into a 5-lobed tube. The petals are lobed at the tip and have appendages at point where the broader upper portion (blade) joins the narrower lower segment (claw). The various species are found in a variety of habitats.

Sleepy Silene (*S. antirrhina*) This annual occurs in open areas, usually associated with burns up to 6,000 feet in the southern California mountains.

Silene (*S. lemmonii*) This slender perennial grows in the open

woods from 3,500 to 8,000 feet. Flowers from June to August.

Silene (*S. parishii*) This perennial occurs on dry gravelly slopes or in rocky places from 6,000 to 11,000 feet in the San Bernardino mountains. Flowers from July to August.

Bladder Campion (*S. vulgaris= S. cucubalus*) Bladder Campion grows along roadsides and gravelly riverbanks at low elevations.

Ecology & Ethnobotany: The young shoots of Bladder Campion can be used as potherbs. The sap of *S. antirrhina* was used by Miwok Indians in California to paint designs on the faces of young girls. The designs were cosmetic, not ritualistic.

Starwort, Chickweeds (*Stellaria*)

Description: There are mostly low, annual or perennial herbs with flowers in an open inflorescence in the leaf axils or at the ends of stems. The 5 sepals are separate to the base. Petals are white and deeply lobed or lacking.

Chickweed (*S. calycantha*) This prostrate to erect perennial occurs on mossy banks, bogs, dry creeks, wet meadows, and shaded areas from 5,000 to 11,000 feet in the Tehachapi and San Gabriel mountains northward.

Starwort (*S. crispa*) Occurs in moist places and meadows between 7,000 to 9,200 feet in the San Jacinto and San Bernardino mountains. Flowers May to August.

Chickweed (*S. jamesiana* = *Pseudostellaria j.*) This species occurs in the San Bernardino mountain between 7,000 to 8,000 feet, as well as Frazier and the Tehachapis. Blooms May to August.

Starwort (*S. longipes*) This tufted perennial grows up to 10 inches tall. The leaves are opposite, lanceolate and rigid. The flowers are terminal, solitary or in few-flowered clusters. Starwort is fairly common in moist places from 4,500 to 10,500 feet. Flowers from May to August.

Common Chickweed (*S. media*) This species that may occur in gardens and lawns in the resort areas. Blooms from February to September.

Shining Chickweed (*S. nitens*) This annual occurs on sand dunes, streambanks, open woodlands, beneath boulders, and in disturbed areas below 5,000 feet in southern California.

Ecology & Ethnobotany: While the uses of other Starwort are unknown, the young shoots of *Stellaria media* have been used as salad herbs or potherbs if cooked like spinach. Although it is edible raw, we prefer to boil for a few minutes before eating. Since the plants are usually quite small and only the youngest parts are good, Chickweed can be tedious to collect. The greens are low in calories and packed with copper, iron, phosphorus, calcium, potassium, and Vitamin C - valued in the prevention and treatment of scurvy.

Medicinally, *S. media* can be used as a tonic, in large quantities a laxative, and diuretic. For itchy skin, make a strong tea and wash the area. A poultice of the plant has been used to treat skin sores, ulcers, and infections as well as eye infections and hemorrhoids.

STAFF-TREE FAMILY (Celastraceae)

These are shrubs with small inconspicuous flowers borne in the axils of the leaves. The 4-5 sepals are united at the base, and the petals are separate. The fruit is a capsule. There are 60 genera and 850 species distributed in tropical and temperate regions of the world. Ten genera are native to the United States. Several species are cultivated as ornamentals.

Western Burning Bush (*Euonymus occidentalis*)

Description: This is a straggly shrub up to 18 feet tall with smooth, greenish-white branchlets and opposite, deciduous leaves. Leaves are glabrous, ovate, with toothed margins. Flowers are small and occur in 1 to 5 flowered clusters. Petals are purplish-brown and about 1/8-inch long. Western Burning Bush is occasionally found in canyons from Mt. San Jacinto to the mountains in San

Diego County from 4,500 to 6,500 feet. Blooms from April to June.

Ecology & Ethnobotany: Related species were medicinally by various Native Americans.

HORNWORT FAMILY (Ceratophyllaceae)

The Hornwort Family has only one genus, *Ceratophyllum*, with approximately three species distributed worldwide. This aquatic herb occurs in lakes, ponds, and slow streams, and have no roots.

Coon's Tail (*Ceratophyllum demersum*)

Description, Ecology, & Ethnobotany: This is a rootless, submersed or free-floating aquatic forb with slender, lax, and much branched stems. The sessile leaves are in whorls of 5-12, and the blades are dissected into linear, filamentous segments whose shape varies with the position on the plant. The minute flowers have no petals, and are borne in the axils of the leaves. Coon's Tail is a common plant in standing or slowly flowing water of rivers, sloughs, and ponds to about 7,000 feet. Flowers June to August. The Maidu Indians used Coon's Tail to

 make a soothing lotion that was used on sore or inflamed skin.

GOOSEFOOT FAMILY (Chenopodiaceae)

Approximately 102 genera and 1,500 species in the Goosefoot Family are found worldwide. Fourteen genera are native to the United States, mostly in the West. This family includes several food plants (e.g., beets and spinach) and weeds (e.g., Russian Thistle).

Four-wing Saltbush (*Atriplex canescens*)

Description: This erect, branched shrub grows to 7 feet tall. It has gray, scurfy, and scaly foliage. Leaves are linear to oblong and entire. Flowers are unisexual with separate male and female plants. Male flowers occur in clusters of spikes, forming a long terminal panicle. The fruit is a small, butterfly-like flattened disk with a hard body and 4 wings about ½ inch across. Wingscale Saltbush is fairly common on dry slopes below 7,000 feet. It is particularly common in the deserts. Flowers from June to August.

Ecology & Ethnobotany: There are many uses for these plants, from food to medicine and dyes, as well as soap and spice. The young leaves of many species can be cooked and eaten as greens and have a very distinct salty taste. We've often added them to otherwise bland foods to make our wild meals less boring. Add the leaves to meats while cooking will help spice them up. The seeds were parched, ground into flour,

and made into mush. They can also be soaked in water for a few minutes to make a rather pleasant tasting drink. The Navajo used the flowers to make puddings. The ashes of *A. canescens* (Fourwing Saltbush) make a good substitute for baking soda.

Medicinally, the various species provided the Native Americans with many uses. As an analgesic, the Navajo used the leaves of *A. argentea* (Silverscale Saltbush) as a fumigant for pain, and the Zuni made a poultice from the chewed root for application to sores and wounds. A warm poultice made from the pulverized root Fourwing Saltbush was used to treat toothaches.

The leaves and roots of many species were used as a soap. They were rubbed in water for lather and used in washing clothing and baskets. Many Native Americans also carved arrowheads from the wood for use as weapons and for hunting. The seeds of some species were also used in making a black dye.

Goosefoot, Lamb's Quarters (*Chenopodium fremontii*)

Description: The species in this genus are annuals with mealy or glandular foliage. Leaves are alternate and entire, toothed, or lobed. Flowers are borne in dense clusters in the leaf axils or terminal inflorescence. There are 2 to 5 green or red sepals. The genus name comes from *Cheno*, meaning goose, and *podium*, meaning foot, because the triangular leaves resemble the shape of a goose's foot. Oil of chenopodium, distilled from the fruits, contains a broad-spectrum vermifuge that is widely used in veterinary medicine.

Lamb's Quarters (*C. album*) This is a common weed that occurs below 6,000 feet in a wide variety of habitats.

Pigweed (*C. atrovirens*) This annual plant is found in open, dry places and coniferous forests between 4,000 to 11,000 feet. It has also been observed growing near Baldwin Lake in the San Bernardino mountains.

Pitseed Goosefoot (*C. berlandieri*) This annual is quite common in open, often disturbed areas up to 7,000 feet in the southern California mountains.

California Pigweed (*C. californicum*) This is common on dry slopes below 5,500 feet. Blooms March to June.

Goosefoot (*C. chenopodioides*) This species occurs in alkaline places above 5,000 feet in the San Bernardino mountains. Flowers August to October.

Pigweed (*C. desiocatum*) This species occurs in dry places between 4,000 to 11,000 feet in the montane forests of San Jacinto and San Bernardino mountains and northward.

Fremont's Pigweed (*C. fremontii*) This erect pale green annual common in dry places from 5,000 to 8,500 feet. Flowers from June to October.

Inconspicuous Goosefoot (*C. incognitum*) This plant is found in dry places between 6,000 to 8,000 feet in the montane forests from San Jacinto mountains and north. Flowers July to August.

Goosefoot (*C. leptophyllum*) This species occurs in dry places between 5,000 to 8,000 feet in sagebrush, pinyon/juniper, and yellow pine forest in the San Bernardino mountains northward. Flowers July to September.

Ecology & Ethnobotany: Leaves, tops, and seeds of all species can be used as an emergency or basic food and are quite tasty and nutritious. High in protein, the greens are a good source of Vitamins A and C, iron, potassium, and are extremely rich in calcium. Since it does not become bitter with age, both young and old plants can be used. Leaves may be used raw in salads or boiled in water like spinach. The water can be saved and used as a yellow dye. The leaves were also eaten to treat stomachaches and prevent scurvy. A leaf poultice was used on burns. The flower buds and flowers can be used as potherbs. A single plant can produce up to 70,000 seeds. Seeds can be ground as flour for use in bread or cooked as mush. Seeds can also be eaten without grinding or incorporated into pinole (flour made from a mixture of seeds of small plants). The seeds contain about 15% protein and 55% carbohydrates, more than is found in corn. The seeds can also be used as a coffee substitute.

Large quantities of the plant should not be eaten as many species contain high levels oxalic acid which tends to bind calcium and prevent its proper absorption into the body. Additionally, *Chenopodium* has been known to accumulate toxic levels of nitrates and may cause livestock poisoning. But, because large quantities of the plant must be consumed to cause problems, this type of poisoning may be unlikely. Cooking or freezing of *Chenopodium* apparently breaks down the oxalic acid.

The hard root of some *Chenopodium* species was stored until needed, then grated on a rock to make soap. The leaves were also used to make soap, but are not as effective as the roots.

Warning: *Chenopodium* greens contain oxalic acid and its salts, which can reduce calcium absorption if eaten in large quantities.

Kochia (*Kochia*)

Description: Three species of Kochia may be found in southern California, but it is *K. scoparia* (Common Kochia) that is most commonly encountered. Common Kochia is a bushy annual with stems up to three feet tall. The leaves are alternate, narrowly lance-shaped and tapered at both ends. The herbage may or may not be covered with hairs. Flowers are solitary or in clusters in spikes. The species is common in open, disturbed habitats at low elevations.

Ecology & Ethnobotany: Common Kochia is native to Europe and was introduced into the United States as an ornamental. It has since escaped and has become well established. In Asia, Japan, and China, Common Kochia was cultivated for its seeds. The tips of the young shoots can be prepared as potherbs. The seeds can be eaten raw or cooked, or ground into meal and used in bread making.

Winterfat (*Krascheninnikovia lanata*)

Description: In many older references, Winterfat is also known as *Eurotia lanata*. Winterfat is a small shrub found at the lower plains and foothills elevations, often in saline or alkaline areas. The leaves are alternate, narrow and entire, whereas the flowers occur in heads or spikes in the axils of the leaves.

Ecology & Ethnobotany: While the edibility of this species is unknown, it is considered an important forage plant for horses and other livestock. Medicinally, the plant has been used by many Native American tribes. For example, the Hopi Indians used the powdered root for burns, and a decoction of the leaves was used for fevers. The Navajo made a poultice of the chewed leaves and applied it to a poison ivy rash. The Navajo also incorporated the stems and leaves of this plant in sweat house ceremonies by placing them on hot rocks for the Mountain Chant.

Povertyweed (*Monolepis nuttalliana*)

Description: Povertyweed is a low growing winter annual with prostrate or ascending stems. The leaves are somewhat succulent and lance-shaped, broadened and lobed at the base. Flowers are borne in dense clusters at the leaf bases and the solitary sepal is reddish in color. The seeds are dark brown. The plant is found in open disturbed habitats at the lower elevations.

Uses: The above ground parts of Povertyweed may be eaten as a potherb. The seeds are also edible. Another species reported to occur in southern California is *M. spathulata* (Beaver Monolepis), but its uses are unknown.

Prickly Russian Thistle, Tumbleweed (*Salsola tragus*)

Description: This is not a true thistle (*Cirsium*), but a many branched annual with purplish striped stems up to three feet tall in a rounded form. The lower leaves are threadlike; the upper leaves are awl-like and spine-tipped. The plant may or may not be hairy. When mature, the whole plant becomes rigid, breaks off at ground level, and becomes a "tumbleweed" blowing across the open plain. Flowers are solitary in the leaf axils and are subtended by spiny bracts. Russian Thistle is common in open, disturbed habitats, particularly around agricultural areas at low elevations. It was introduced to the United States from Europe. Fortunately, Russian Thistle is not an aggressive competitor and does not appear to replace native plant species. However, it is still considered a noxious weed because of its distributional pattern and spines.

Ecology & Ethnobotany: This unsavory looking plant is edible. The young parts of the plant may be boiled and eaten as a potherb or chopped raw into a salad. On older plants, clip the tender branch tips that are green. We find the taste of the plant greatly improves when cooked in butter and lemon. In Europe, the ashes of the plant were once used in the production of carbonate of soda known as Barilla.

Warning: The older parts of the plants contain significant quantities of nitrates and oxalates and may be toxic if eaten in quantity.

MORNING GLORY FAMILY (Convolvulaceae)

Members of this family are herbs, shrubs, or trees. In some species, a milky latex is present. The flowers are usually 5-merous with five united petals. There are approximately 50 genera and 1,400 to 1,700 species in this family, distributed in tropical and temperate regions. Nine genera of native to the United States. The family is of some economic importance because of the sweet potato (*Ipomoea batadas*), several weeds, and ornamentals.

Sierra Morning Glory (*Convolvulus arvensis*)

Description: This twinning and trailing perennial with grayish, woolly herbage grows to 12 inches long. The leaves are triangular-shaped and woolly on both sides. The flower is cream white. This species grows on dry slopes and ridges from 1,500 to 6,000 feet in Ventura County northward. Flowers from June to August. The genus name is Latin, *convolvere*, meaning to entwine.

Ecology & Ethnobotany: This species was used by Native Americans as a cold leaf tea as a wash on spider bites. A tea from the flowers was used for fevers and wounds. A tea was also made from the leaves and stems

by the Kashaya and Pomo women to stop excessive menstruation (Strike 1994). In European folk use, the flower, leaf, and root teas were considered a laxative. The root is considered to be a strong purgative, cathartic, and diuretic. The powdered root stock was used as a laxative in ancient and modern China.

DOGWOOD FAMILY (Cornaceae)

There are 12 genera and 100 species in the Dogwood Family including many ornamentals. A single genus, *Cornus*, occurs in the United States. Dogwoods are trees or shrubs, often with tiny flowers surrounded by petal-like bracts which resemble a single large flower. The leaves are opposite and simple.

Dogwood (*Cornus*)

Description: Dogwoods are shrubs or semi-woody perennials with simple leaves that are opposite or whorled. The flowers mature into red or white drupes. The species can be found in moist mountain and foothill forests, preferring partial shade, up to the subalpine zone.

Mountain Dogwood (*C. nuttallii*) This tree has small flowers that are greenish and occur in a head that is subtended by 4 to 7 white or pinkish petal-like bracts. The actual petals of the flowers are less than 1/8-inch long. Mountain Dogwood is found in woods below 6,000 feet from San Diego to Los Angeles County. Blooms April to July.

Western Red Dogwood (*C. occidentalis*) This shrub has white flowers that occur in a dome or flat-topped cluster that is up to $2\frac{1}{2}$ inches wide. Western Red Dogwood grows in moist places, generally below 8,000 feet. Flowers from May to July.

Red Osier Dogwood (*C. sericea*) This shrub occurs in many moist habitats below 8,000 feet. The species is highly variable with many local forms.

Ecology & Ethnobotany: The berries of *C. nuttallii* (Mountain Dogwood) can be eaten raw or cooked. Strike (1994) suggests that the fruit contains enough protein, carbohydrate, and fat to sustain life when other food sources are not available. The inner bark of Mountain Dogwood twigs can be scraped off and used as an additive to tobacco mixes for smoking. Pounded twigs can be used as a toothbrush. The wood for both species was used for bows and arrows, fishing hooks, and other implements. The bark was boiled and used to make a brown dye.

The fruits of a related species, *C. canadensis* (Bunchberry Dogwood), may be eaten raw or cooked. Since the berries are rather bland, we like to mix them with other better tasting berries. The unripe berries may cause stomachaches or act as a laxative. The chewed berries have been used as a poultice to treat local burns. A cold and fever remedy can be made by boiling dried root or bark (the root is more potent). The feathered bark can be used as a toothbrush. Fresh bark is a cathartic. Leaf tea was used for aches and pains, kidney and lung ailments, coughs, and fevers, and as an eye wash. Dogwood has earned a reputation as an anti-inflammatory and general analgesic due to the presence of cornine and other flavonoid compounds. Researchers are studying these properties as an anti-cancer agent. The current interest by pharmaceutical companies may stem from the fact that Native Americans used Dogwood as an antidote for a variety of poisons.

The berries of *C. sericea* (Red Osier Dogwood) were sometimes consumed by Native Americans. We have found them to be extremely bitter. In fact, in large quantities they may be toxic. The Blackfeet Indians used the bark of Red Osier Dogwood as a laxative. Other Native Americans smoked the inner bark as part of a ceremonial herb blend.

STONECROP FAMILY (Crassulaceae)

Members of this family are succulent herbs or shrubs. There are 35 genera and over 1,500 species worldwide, of which nine genera are native to the United States. They are of no real economic importance except as ornamentals.

"Quick Key" to the Stonecrop Family

1. Petals distinct to the base; stems green - *Sedum*
1. Petals fused at the base; corolla tubular; stems red – *Dudleya*

Live-forever (*Dudleya*)

Description: These are fleshy perennials with leaves in basal rosettes. The genus is named after W.R.

Dudley, a western United States botanist that live from 1849-1911.

Live-forever (D. cymosa) This glabrous and pale perennial plant with bright yellow to orange colored flowers grows on dry rocky places between 2,000 to 8,500 feet in San Bernardino County northwards. Flowers from April to July.

Live-forever (D. saxosa) This pale green, glabrous perennial with yellow flowers that dry red, grows on dry, stony slopes from 800 to 5,500 feet in the desert mountains. Flowers from April to June.

Ecology & Ethnobotany: The stems and laves of many *Dudleya* species were eaten. *D. saxosa* was considered by the Cahuilla as a delicacy as the fleshy leaves were eaten raw.

Heated *Dudleya* leaves were used as to poultice and remove calluses. A leaf poultice of *D. cymosa* hastened the healing of sores or wounds. It was also used as a diuretic.

Stonecrop (Sedum)

Description: The Latin word *sedere* means "to sit," possibly referring to the tendency of many species growing low to the ground. Stonecrops are well-adapted to survival in shallow soil or on rocky outcroppings. The succulent leaves and stems have a waxy coating to help reduce water loss. The reddish color of the foliage in some species is enhanced by sunlight and occurs most often in plants in hot exposed sites.

"Quick Key" to the Stonecrops

1. Stem leaves numerous, about the same as the basal leaves; flowers white - *S. niveum*
1. Basal leaves in rosettes, stem leaves few and reduced; flowers yellow - *S. spathulifolium*

Davidson's Stonecrop (*S. niveum*) This glabrous perennial with white flowers grows in rocky, shady areas between 7,000 to 9,600 feet in the San Bernardino and Santa Rosa Mountains. Flowers from June to July.

Sedum (*S. spathulifolium*) This glabrous perennial with yellow flowers grows in rocky places between 2,500 to 7,000 feet in the San Bernardino and San Gabriel Mountains. Flowers in June and July.

Ecology & Ethnobotany: The young leaves and stems of all species can be eaten as a salad or boiled as a potherb. We find them slightly tart and crisp - a wonderful addition to salads and trail snack. However, some species have emetic and cathartic properties, and can cause headaches. In an emergency, Stonecrop can be eaten raw to allay hunger and thirst. The plants are best when collected before flowering since they tend to become bitter and fibrous in late summer. The green fleshy leaves are high in Vitamins A and C. The tubers can also be boiled and eaten.

Sedum has also been reported as being slightly astringent and mucilaginous. It is valuable in the treatment of wounds, ulcers, lung disorders, and diarrhea. As a field remedy for minor burns, insect bites, and other skin irritations, just squeeze the juice onto the affected area. Decoctions of the plant were also used for sore throats and colds, and as an eye wash.

CROSSOSOMA FAMILY (Crossosomataceae)

Members of this family are shrubs with simple, entire, and coriaceous leaves. The flowers are white or sometimes rose-tinged, and the fruit (follicle) is green at maturity. There are 3 genera and about 8 species in the western United States and Mexico.

Nevada Greasewood (*Glossopetalion spinescens*)

Description: This shrub has stems that are more/less spiny and yellowish in age. The oblong to lanceolate shaped leaves have pubescent surfaces. This species occurs from 3,500 to 7,000 feet at the northern portion of the San Bernardino mountains. Flowers from April to May.

Ecology & Ethnobotany: The Shoshoni made a decoction of the shrub that was used for tuberculosis.

CUCUMBER OR GOURD FAMILY (Cucurbitaceae)

Members of this family are annual or perennial herbs that are climbing or prostrate, with spirally coiled tendrils. The leaves are alternate, often palmately lobed. The fruit is a berry (often referred to as a pepo) with a leathery or hard exocarp. There are approximately 100 genera and 850 species distributed in the warmer regions of the Old and New World. Fourteen genera are native to the United States. The family is economically important as a source of many food plants and ornamentals.

California Man-root (*Marah fabaceus*)

Description: These are climbing perennials from swollen, woody roots. The leaves are palmately lobed, and heart-shaped at the base. The flowers are white and bell-shaped. The fruits are football-shaped, fleshy, and with weak or soft spines. They tend to spread along the ground and climb onto other vegetation. The genus name is from the Latin, meaning bitter, referring to the taste of all plant parts.

Ecology & Ethnobotany: Strike (1994) indicates that *Marah* seeds were roasted and eaten by several Native American tribes. The seeds are supposedly high in protein and oil, were crushed into a flour. The leaves may have been eaten by the Chumash. As a warning, however, consider the plant **poisonous**.

Marah roots were brewed into a tea and used as a strong purgative by the Luiseno. California Man-root was used by the Pomo as a dermatological aid. They pounded the nuts and grease, and rubbed the concoction on the head for falling hair.

Finally, *Marah* roots are poisonous, and they were mashed and put into streams to stupify fish and make them easier to catch.

DODDER FAMILY (Cuscutaceae)

Members of the Dodder Family are leafless, rootless, parasitic herbs that lack chlorophyll. The stems are thread-like and often yellowish in color. The small flowers have 4 or 5 distinct sepals, and 4 or 5 united petals. The fruit is a dry or fleshy globose capsule. The family has one genus (*Cuscuta*) with approximately 170 species. The genus is native to the United States and may cause great losses to crop plants. The family was once included in the Convolvulaceae (Morning Glory Family).

Dodder (*Cuscuta*)

Description: Dodders are leafless, twining perennials with slender stems that colored pink, whitish, or yellowish, never green. Both the leaves and pink to white flowers are highly reduced. Dodder can normally be identified only with a microscope or hand lens. The many species of *Cuscuta* parasitize different flowering plant hosts at low elevations.

Dodders have a very unique life cycle. The small seeds usually germinate in the soil and produce slender stems without seed leaves (cotyledons). Unless the slowly rotating plant encounters a host plant within a short period of time, the Dodder seedling will wither and die. However, if the seedling encounters the living stem of a susceptible host plant, the Dodder will twine around it and at certain points develop suckers that penetrate the tissue of the host. Nutrition is received through these suckers. Dodder then losses all contact with the soil. After a period of growth, small flowers develop and large amounts of seeds are produced to start the process all over again (Franklin and Mulligan 1987).

Ecology & Ethnobotany: Dodder was often called "love vine" and "vegetable spaghetti" by some Native Americans, but they are generally not considered to be edible and may cause digestive upset. However, the seeds of *C. californica* (Chaparral Dodder) were parched and eaten by Maidu Indians in California.

Chaparral Dodder, when brewed as a tea, was considered to be an antidote for black widow bites. However, only the Dodder from *Eriogonum fasciculatum* (California Buckwheat) was used for this purpose. Other Native Americans chewed a mass of Dodder and stuffed it in their nose or pulverized the plant and sniffed the powder to stop nosebleeds.

Dodder stems were used by the Cherokees as a poultice for bruises. In China, the stems of some species of Dodder are used in lotions for inflamed eyes. Moore (1979) indicates that a rounded teaspoon of the chopped Dodder is a good laxative-cathartic. In smaller quantities, and drunk every few hours, it is said that it will aid in spleen inflammations, lymph node swellings, and "liver torpor." Additionally, handfuls of Dodder can be gathered and used as scouring pads for cleaning.

DURANGO ROOT FAMILY (Datiscaceae)

This family is comprised of three genera with a total of 4 species. A single species of *Datisca* occurs in California and ranges into Mexico. The other members of this family occur in eastern Asia and on the adjacent Pacific islands.

Durango Root (*Datisca glomerata*)

Description: This is a glabrous erect perennial plant with ovate lanceolate leaves that are pinnately divided into lanceolate toothed segments. Flowers occur in axillary clusters and there are no petals. Durango Root grows in dry stream beds and washes below 6,500 feet. Flowers from May to July.

Ecology & Ethnobotany: The pulverized root was used as a decoction to wash sores and heal rheumatic pain. The Costanoan Natives used a decoction of the plant to relieve sore throats and swollen tonsils.

Warning: All parts of Durango Root should be considered toxic.

TEASEL FAMILY (Dipsacaceae)

These are mostly herbaceous plants with opposite leaves. There are approximately 10 genera and 270 species found mostly in the Old World. None are native to the United States, although *Dipsacus* is widely naturalized and has become quite weedy.

Teasel (*Dipsacus fullonum*)

Description: This is a stout, prickly tap-rooted plant up to six feet tall. The leaves are opposite and lance-shaped and distinctly prickly on the lower surface of the midrib. The small, bluish-purple flowers occur in a large, terminal flower head that is egg-shaped and armed with numerous, sharp-pointed bracts. This non-native plant from Europe is found throughout North America on disturbed soils with appreciable water holding capacity. The plant can be opportunistic, and displaces desirable native plants. The genus name is from Greek *dipsa*, which means thirst, and refers to the accumulation of water in the cup-like bases of the joined leaves. It is not uncommon to find insects and other invertebrates living in the water found in the leaves.

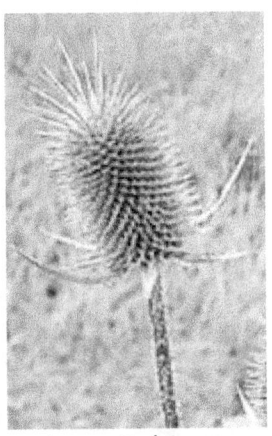

Ecology & Ethnobotany: The dried flower spike of Teasel looks like a caged "thistle" and is sometimes mistaken for Thistle (*Cirsium*) because of its "thorny" character and bluish flowers. However, Thistle is a member of the Sunflower Family (Asteraceae) and has alternate, not opposite leaves.

While there are no documented edible nor medicinal uses for the plant, they are sought after for use in dry flower arrangements. The dried inflorescence is often sprayed with some interesting colors and sold in craft stores.

Cloth cleaners use the dried flower heads to remove the nap from wool or cloth after beating and cleaning. Apparently, the Teasel heads perform the task so well that man-made tools have not replaced them.

OLEASTER FAMILY (Elaeagnaceae)

Plants in this family are shrubs or trees with alternate or opposite, and silvery-gray leaves. The fruits are drupe- or berry-like. There are three genera and approximately 45 species in this family distributed in North America, south Europe, Asia, and eastern Australia.

Buffaloberry, Soapberry (*Shepherdia argentea*)

Description: Buffaloberry is a low spreading shrub with opposite leaves, each with a dark green upper surface and lighter underside that is covered with tiny, brown scales. The inconspicuous flowers are yellow-green; the fruits range in color from yellow to bright red. Buffaloberry grows along streams from 3,500 to 6,500 feet. Blooms April to May. It is also found in recently burned areas. The genus is named after John Shepherd (1764-1836), curator of the Liverpool Botanic Garden.

Ecology & Ethnobotany: Another common name for this species is Soapberry. The berries contain a significant amount of saponin which not only gives the plant its bitter taste, but also whips up into a frothy mass called "Indian Ice Cream."

Native Americans used the berries of these plants extensively, both fresh and dried. The berries of this species are at first pleasant, then the soap-like bitterness prevails. We enjoy cooking them with sweeter tasting berries such as Thimbleberries and Serviceberries, and a large amount of sugar. We found the berries to be somewhat unattractive for general use, but a valuable consideration in emergencies. The berries taste better after a few good frosts during the fall. They can also be used in the making of pemmican or jelly. Dried into cakes, the berries can be stored for winter.

Infusions of the stems and leaves were drunk as a tonic beverage. The berries can be crushed and made into a tea for use as a liquid soap. Native Americans used the tea to relieve constipation. Hart (1976) indicates that the Flathead and Kootenai Indians made solutions from the bark of Buffaloberry for eye troubles.

"Indian Ice Cream"

To make Indian Ice Cream, Native Americans placed a small number of berries into a bowl with a little water, then used a special stick with some grass tied on one end to beat the fruit. The result was foamy concoction. In recent times, sugar was added to improve the taste. Care must also be taken in picking and preparing the berries so that they do not come in contact with oil or grease of any kind, or they will not whip. Indian Ice Cream is still served in many households, especially at parties and family gatherings in the Pacific Northwest.

HEATH FAMILY (Ericaceae)

There are about 50 genera and 2,500 species in the Heath Family, the majority occurring in acidic soils. In the United States approximately 25 genera are indigenous. Economic products provided by this family include food plants, oil of wintergreen, and many ornamentals. Some herbaceous members are mycotrophic, that is they depend on fungi for nutrient uptake and lack chlorophyll. The list of genera and species here includes parasitic (families Monotropaceae and Pyrolaceae) and hemiparasitic herbs in addition to the autotrophic plants.

"Quick Key" to the Heath Family

1. Plants not green; leaves scale-like - **2**
1. Plants green; leaves expanded - **3**

2. Plant pinkish, but becoming brown with age and tall and slender - *Pterospora*
2. Plant red, short and stout - *Sarcodes*

3. Plant more/less herbaceous; leaves basal or on short stems - **4**
3. Plant woody; leaves occurring on stem - **5**

4. Flowers 1 to few in head - *Chimaphila*
4. Flowers few to many - *Pyrola*

5. Fruit a drupe or berry; plants generally of lower elevations or dry habitats - **6**
5. Fruit a capsule; plants of higher elevations or moist habitats - **7**

6. Ovary and fruit smooth; fruit not juicy - *Arctostaphylos*
6. Ovary and fruit with small rounded protuberances; fruit more/less juicy - *Arbutus*

7. Leaf greater than $\frac{1}{4}$-inch wide and leaf margin not strongly rolled under - *Rhododendron*
7. Leaf less than $\frac{1}{4}$-inch wide, leaf margin strongly rolled under - *Phyllodoce*

Pacific Madrone (*Arbutus menziesii*)

Description: This evergreen tree grows up to 120 feet tall, and has peeling bark and stems that are red and appear polished. The large leaves are alternate, leathery, entire or toothed, ovate, and glabrous. They are dark green above and paler below. The flowers are urn-shaped, white to pink in color, and occur in terminal clusters. The fruit is a globose, red-orange berry. It has a rough, granular surface. Madrone grows in wooded, moist areas below 5,000 feet in scattered localities in southern California; more abundant north of Santa Barbara. Blooms from March to May.

Ecology & Ethnobotany: The ripe berries can be eaten raw or cooked. Some Native Americans steamed the berries in a basket with Madrone leaves on top, then dried and stored the berries. To eat, these then were soaked in warm water to soften. The berries do not store well as they will decay when bruised. We have found them best when mixed and crushed with Manzanita berries. The berries can also be used in making a tea. Simply crush the berries in water like you would Manzanita.

Caution: If eaten in large quantities, the berries could cause a person to vomit

Manzanita (*Arctostaphylos*)

Description: The genus name means "bear grape" and refers to the fondness shown by bears for the fruits of these shrubs, many of which are known as Bearberry.

"Quick Key" to the Manzanitas

1. Flowers subtended by pink bracts; leaves rough to the touch - *A. pringlei*
1. Flowers not subtended by pink bracts; leaves smooth, glabrous - **2**

2. Berry sticky, glandular - **3**
2. Berry glabrous - **4**

3. Low spreading shrub, 2 to 4 feet tall - *A. glandulosa*
3. Erect shrub or tree, 6 to 26 feet tall - *A. glauca*

4. Erect shrub, 6 to 9 feet tall - *A. pungens*
4. Low shrub with branches spreading along ground - **5**

5. Leaves almost roundish; $\frac{3}{4}$ - $1\frac{1}{2}$ inches wide - *A. patula*
5. Leaves ovate; $\frac{1}{2}$ to $\frac{3}{4}$-inch wide - *A. parryana*

 Eastwood Manzanita (*A. glandulosa*) This low, spreading shrub up to 4 feet tall occurs at lower elevations in the mountains. Flowers from February to March.

 Bigberry Manzanita (*A. glauca*) This erect shrub or tree occurs on dry mountain slopes at lower elevations. Flowers from February to April.

 Parry Manzanita (*A. parryana*) This low, spreading shrub with ovate leaves occurs on dry slopes from 4,000 to 7,500 feet. Flowers from February to March.

 Greenleaf Manzanita (*A. patula*) This low shrub with almost roundish leaves is found from Kern

County northward between 2,000 to 9,000 feet. Flowers May to June.

Pink-bracted Manzanita (*A. pringlei*) This shrub with bracts subtending the flowers grows on dry slopes in the mountains between 4,000 to 7,500 feet. Flowers February to March.

Mexican Manzanita (*A. pungens*) This erect shrub can be found on dry slopes between 3,000 and 7,000 feet in the mountains. Flowers from February to March.

Ecology & Ethnobotany: The berries of all Arctostaphylos are edible. They may be eaten raw, and it is suggested that they not be eaten in large quantities since they may be hard to digest. Constipation or indigestion are common maladies of eating too much. The berries can also be stewed, or dried and ground into meal and cooked as mush. A cider can also be made from the berries. The seeds alone can be collected and ground into meal too.

The leaves can also be boiled in water, allowed to cool, and the decoction applied to stop the itching and spread of Poison Oak. The internal consumption of the leaf tea often results in urine becoming alkaline and bright green. This is caused by the urinary antiseptic hydroquinolone, and it is relatively harmless. These hydroquinones (particularly arbutin) are

strongly antibacterial and are effective against *Klebsiella* and *E. coli*, which are often associated with urinary infections.

Many species were mixed with tobacco and smoked. The general effect of some Native American tribes smoking the leaves of Manzanita was intoxication due to its narcotic content. Leaves are an astringent due to the tannic acid and have been used to tan hides. The leaves can also be chewed to stimulate saliva, particularly when one is thirsty.

Manzanita Cider

To make manzanita cider, simply crush the ripe or green berries in a container, then pour an equal volume of scalding water over them. After the mixture has cooled and the solids have settled, decant the liquid and drink. The cider will be a little dry to the taste.

Pipsissewa (*Chimaphila umbellata*)

Description: This is a short evergreen semi-shrub (woody only at the base) that originates from a long creeping rootstock. The leaves are whorled and leathery. Pipsissewa grows in dry, shady forests from 1,000 to 10,000 feet in the San Jacinto Mountains and north. Flowers from June to August. Another species, *C. menziesii* may also be found in the southern California mountains.

Ecology & Ethnobotany: The roots and leaves of Pipsissewa may be boiled, and the liquid cooled for a refreshing drink that is high in Vitamin C. The leaves may also be nibbled raw, but because of their astringency and tough texture we found them unappealing.

Pipsissewa was an important herb to Native Americans for treating rheumatism. A tea from the leaves was used for the purpose to treat rheumatism and kidney problems. The plant contains quinone glycosides, such as that found in *Arctostaphylos*, but is less astringent and more a diuretic, making it better for long-term use. The plant was also mixed with tobacco for smoking. Pipsissewa produces a natural antibiotic that can be used by humans. Hot infusions of Pipsissewa can be taken to induce perspiration in the treatment of typhus, and the berries can be eaten for stomach disorders.

Pipsissewa is a "secret ingredient" in certain popular soft drinks. In the Pacific Northwest, these plants, as well as certain species of *Pyrola* are under commercial harvesting pressure and may be slowly disappearing.

Red Mountain Heather (*Phyllodoce breweri*)

Description: This is a low shrub, 1 foot or less tall, with leaves that are alternate, crowded, linear, and needle-like with rolled-in margins. The flowers occur in a terminal raceme and are bright rose-purple. Corolla is bell-shaped and hangs downward. Red Mountain Heather occurs in the higher elevations in moist areas and alpine meadows from 6,000 to 12,000 feet. Blooms in the early summer.

Ecology & Ethnobotany: A related species, *P. empetriformis* (Pink Mountainheath), was used by the Thompson Indians (southwestern British Columbia, Canada) as a tuberculosis remedy. Apparently, a decoction of the plant was taken over a period of time for tuberculosis and spitting up blood.

Woodland Pinedrops (*Pterospora andromedea*)

Description: This is a brownish-red plant with sticky stems up to 3 feet tall with pale-yellow flowers. Found in deep humus of coniferous forests between 2,500 to 8,500 feet, usually associated with Ponderosa Pine (*Pinus ponderosa*). Flowers from June to August.

Ecology & Ethnobotany: Foster and Duke (1990) indicate that Native Americans used a cold tea made from the pounded stems and fruits to treat bleeding from the lungs. As a dry powder, the plant was used as a snuff for nosebleeds.

Wintergreen (*Pyrola*)

Description: In general, Wintergreen are low, smooth perennial herbs with shiny, leathery leaves that are clustered at the base. The flowers are waxy and nodding. *Pyrola* stems from *pyrus* for pear, probably since the leaves of many species resemble pear leaves.

"Quick Key" to the Wintergreens

1. Flowers all dangling on one side of the axis - *P. secunda*
1. Flowers not on one side of the axis - *2*

2. Petals pink - *P. asarifolia*
2. Petals cream to slightly greenish - *P. picta*

Pink Wintergreen (*P. asarifolia*) This low perennial plant with pink flowers that grow on both sides of the stem grows in moist, shaded woods from 4,000 to 9,000 feet in the San Bernardino Mountain and northward. Flowers from July to September.

White-veined Wintergreen (*P. picta*) This low perennial with cream or slightly greenish flowers grows in the shaded forest on rich humus, between 3,000 to 9,500 feet. Flowers from June to August.

One-sided Wintergreen (*P. secunda = Orthilia s.*) This low perennial has flowers dangling on one side of the stem and grows in dry shady woods from 3,000 to 10,500 feet in the San Jacinto and San Bernardino Mountains and northwards. Flowers from July to September.

Ecology & Ethnobotany: A tea made from the whole plant was used to treat epileptic seizures in babies. A leaf tea was gargled for sore throats, and canker sores, while a tea from the root was a tonic. A poultice from the mashed leaves was used for tumors, sores, cuts, and to relieve the itch of insect bites. The plant is also an excellent astringent and disinfectant for urinary tract infections. The plants contain ursolic acid and the glycosides arbutin and ericolin, which were used in the treatment of kidney problems and skin eruptions.

Pyrola is also used as an ingredient in popular soft drinks. It is said to be an excellent substitute for *Chimaphila umbellata* (Pipsissewa). In some areas *Pyrola* may be exploited (over harvested) for commercial purposes.

Western Azalea (*Rhododendron occidentale*)

Description: This deciduous shrub with shredding bark and large white flowers grows up to 10 feet tall. The flowers are very showy and fragrant. Western Azalea occurs in moist places in the southern mountains below 7,500 feet. Flowers from May to July.

Ecology & Ethnobotany: The leaves may contain andromedotoxin and should be considered poisonous.

Warning: Many plants in this family contain a poisonous compound called andromedotoxin. If consumed in large concentrations this could be harmful, causing vomiting, illness, and even death.

Snow Plant (*Sarcodes sanguinea*)

Description: This red, fleshy, saprophytic plant with scale-like leaves grows up to 20 inches tall. This plant grows in thick humus of the shady forests between 4,000 to 8,000 feet. Flowers from May to July.

Ecology & Ethnobotany: The fleshy plant is edible when prepared like asparagus. However, because this is a protected and rare plant, only in an emergency should this be even considered. A decoction of the leaves and stems was used to treat ulcerated sores, irritated skin, and toothaches. The decoction was also used as a blood tonic.

California Huckleberry (*Vaccinium ovatum*)

Description: his shrubby plant can grow to more than 3 feet tall. The leaves are persistent and leathery, and the white to pink flowers are bell-shaped. The fruit is a berry. Occurs on dry canyon slopes below 5,000 feet. Flowers March to May.

Ecology & Ethnobotany: In general, all Vaccinium berries can be eaten raw or be dried in the form of cakes for future use. The various species we have sampled range in taste from sweet to tart. Hybridization between the species is known to occur, but the fruits are still edible. The berries have also been used as fish bait since they look very similar to salmon eggs. The leaves can be dried to make a tea. The leaves and berries are high in Vitamin C.

SPURGE FAMILY (Euphorbiaceae)

The Spurge family has about 290 genera and 7,500 species distributed worldwide. Among the valuable products of the family are rubber, castor and tung oils, and tapioca. Most members are poisonous, and have milky sap that will irritate eyes and mouth.

Turkey Mullein (*Eremocarpus setigerus*)

Description: Turkey Mullein is a grayish-green annual with a musky smell. It is usually found in dry, often rocky areas at low elevations. The genus name is from the Greek, *eremos* for solitary and *karpus*, for fruit.

Ecology & Ethnobotany: The herbage of the

plant is poisonous. Strike (1994) indicates that the Costanoan Indians used a decoction of the root for dysentery, while the Kawaiisu used the decoction, internally and/or externally, to relieve headaches and rheumatic pains.

Sweet (1976) says that the plant contains a narcotic and the foliage was used by some aboriginal peoples to stupefy fish and poison their arrow points. It has been suggested that the stellate hairs get into the fish's gills and hold them open, so in time the fish would drown. Moore (1989) on the other hand, indicates that it may be diterpenes that are the main cause of the plant's effects on fish. Fresh leaves were bruised and applied as a counterirritant poultice for internal pain and asthma.

Wood Spurge (*Euphorbia palmeri*)

Description: This is a glabrous perennial 4 to 14 inches tall. It has several erect stems which contain a milky juice. The $\frac{1}{4}$ to $\frac{3}{4}$- inch long leaves are ovate, without petioles, whorled, and with toothless margins. The small flowers are subtended by round-ovate leaves. Wood Spurge is common on dry slopes between 4,000 and 9,000 feet. Flowers from May to August. Ventura County south to San Diego County.

The flowers are borne in a complex structure called a cyathium. This cup-like structure contains several male flowers and a single female flower. The genus name is from the Greek "euphorbion," a plant named after Euphorbos, a celebrated Greek physician of the 1st century B.C.

Ecology & Ethnobotany: *Euphorbia* contains toxic principles that will cause severe poisoning if ingested in quantity. Most species contain carcinogenic, highly irritant, diterpene esters and are strong purgatives. The white sap can cause skin irritations and blisters. The Cahuilla Indians in California used both the native and introduced species as a medicine for reducing fever and as a cure for chicken pox and smallpox. The plant was boiled and the afflicted person was bathed in the decoction.

PEA OR LEGUME FAMILY (Fabaceae)

The Pea family is one of the largest plant families in the world. There are approximately 600 genera and 13,000 species worldwide. Next to the grass family, which produce all our grains and cereals, the Pea Family is the second most economically important group of plants in the world. The beans and peas that we eat for dinner, as well as the traditional peanuts at baseball games are found in this family. But before taking a bite of the next legume you see, be aware that the family also contains a number of highly toxic members. The various species of Locoweeds and Milkvetches (*Oxytropis* and *Astragalus*) have caused much loss of livestock.

The pea flower is referred to as "papilionaceous", and means butterfly-like. The flowers are bilaterally symmetrical, consisting of 5 petals. The largest upper petal is called the "banner," the two lateral ones are the "wings," and the two lowest ones are fused at the lower margins to form a boat-like structure called the "keel."

Mock Locust (*Amorpha*)

Description: These are deciduous shrubs with gland-dotted and heavy scented foliage. There are about 45 species in North America. The genus name is from the

Greek, *amorphos*, meaning deformed, because of the corolla.

Mock Locust (*A. californica*) This is a shrub with heavy scented foliage and dotted with glands grows on dry, wooded slopes in the Santa Rosa, Santa Ana, and Santa Lucia mountains. Flowers from May to July.

Desert False Indigo (*A. fruticosa*) This shrub grows in scrub, habitat, in canyons and moist areas usually below 5,000 feet. Blooms from May to July.

Ecology & Ethnobotany: Desert False Indigo was used in making mats, rugs, and bedding. The long stems were the foundation for the bedding material. Additionally, the stems were used to make arrows.

Milkvetch (*Astragalus*)

Description: There are many species of Milkvetch. They are perennial herbs with odd-pinnate leaves that have leafy stipules. This is a difficult genus of perhaps 1600 species, making it the largest genus in the Pea Family. The name comes from the ancient Greek name for a plant in the pea family.

"Quick Key" to the Milkvetches

1. Racemes 10- to 35-flowered; flowers cream colored - *A. douglasii*
1. Racemes 2- to 10-flowered; flowers pink to purple - 2

2. Pod thick, woolly, ovoid; flowers 3/4 to 1 inch long, and light colored - *A. purshii*
2. Pod short, villous, oblong; flowers $\frac{3}{4}$-inch long and purple - *A. leucolobus*

Locoweed (*A. douglasii*) This is a low, spreading perennial that forms clumps 16 to 40 inches across is common in grassy places and on dry fields up to 7,000 feet in the San Gabriel and San Bernardino Mountains. Flowers from April to July.

Locoweed (*A. leucolobus*) This is a low, matted perennial with long hairy or woolly herbage and

mostly basal leaves grows in open areas in the San Bernardino and Santa Rosa Mountains from 6,000 to 7,800 feet. Flowers from May to July.

Woolly Pod (*A. purshii*) This low, matted perennial has long hairy or woolly herbage and mostly basal leaves grows on Mt. Pinos and the Tehachapi Mountains and north between 4,000 to 8,000 feet. Flowers from April to June

Ecology & Ethnobotany: Although the roots, pods, and peas of some species were reported to be eaten by American Indians, this genus is not recommended for consumption. All Milkvetches produce either a toxic alkaloid substance or accumulate selenium from the soil or both. Selenium poisoning of livestock has the following characteristics - lethargy, diarrhea, loss of hair, breakage at the base of hoof, excessive urination, difficulty breathing, rapid and weak pulse, and coma. Death usually results from the failure of lungs and heart.

An interesting side benefit has developed from discovery of plants that grow only in selenium soils. Scientists can use the plants to map areas high in selenium for the purpose of mining the valuable element. Some *Astragalus* species also provide good indicators of uranium ore and copper-molybdenum deposits.

Redbud (*Cercis occidentalis*)

Description: This is a deciduous tree or shrub with simple, alternate, and entire leaves. The flowers are lavender in color and occur in clusters or racemes. The species occurs below 5,000 feet in the southern California mountains.

Ecology & Ethnobotany: The pods and seeds were supposedly eaten by the Maidu and Wintum. The Mendocino treated chills and fevers using redbud bark. This was one of the most important basketry plants and was used regularly in making twined and coiled baskets.

Sweetpea (*Lathyrus*)

Description: Sweetpeas are vines, climbing or supporting themselves on other vegetation. The plants have tendrils at the ends of their leaves. Sweetpeas are found in a variety of habitats from the foothills up to the subalpine.

Brush Pea (*L. brownii*) This glabrous perennial grows on dry slopes from 4,000 to 6,000 feet in the

Tehachapi Mountains and northward. Flowers from April to June.

Canyon Pea (*L. vestitus*) This is a pubescent to glabrous perennial grows in dry places below 5,000 feet in the Santa Monica, San Jacinto, and Santa Rosa Mountains. Flowers from April to June.

Ecology & Ethnobotany: Some species of *Lathyrus* have a history of poisoning humans. Kirk (1975) indicates that an exclusive diet of some species from 10 to 30 days can bring on partial or total paralysis, and Willard (1992) suggests avoiding these plants entirely. Strike (1994) says that the greens and raw seeds of *Lathyrus* in California were eaten by Native Americans. Some of the seeds were parched and made into pinole and was stored for winter use. Weedon (1996) says that the fruit of many species are edible in small amounts, but may cause paralysis and several secondary disorders if eaten in large quantities over time.

Caution: It is best to assume that these plants are poisonous if ingested.

Lotus (*Lotus*)

Description: The five species of Lotus are annual or perennial herbs with pinnately compound leaves. The flowers are pea-shaped, yellow or white, often tinged with reddish or purple. They occur in many habitats at various elevations.

Silver-leafed Lotus (*L. argophyllus*) This much branched perennial has silvery to white woolly herbage grows on dry slopes below 5,000 feet. Flowers from April to July.

Silver Lotus (*L. argyraeus*) This is a matted perennial that grows on dry slopes and flats from 3,500 to 8,300 feet in the San Bernardino, San Jacinto, and Santa Rosa Mountains. Flowers from May to August.

Broad-leaves Lotus (*L. crassifolius*) This erect perennial grows on dry banks and flats from 2,000 to 8,000 feet in the southern California mountains. Flowers from May to August.

Sierra Nevada Lotus (*L. nevadensis*) This is a perennial plant with stems that spread along the ground and form mats that are 12 to 32 inches across grows on dry, sandy, and gravelly slopes from 3,500 to 8,500 feet in the mountains of San Diego, Kern, and Inyo Counties. Flowers from May to August.

Narrow-leaved Lotus (*L. oblongifolius*) This is an erect perennial that grows in wet places below 8,500 feet. Flowers from May to September.

Ecology & Ethnobotany: Many of the species are presumed to be poisonous and should be avoided.

Lupine (*Lupinus*)

Description: The many species of Lupine are showy perennial or annual herbs with palmately compound leaves. Flowers are blue, violet, rarely white, rose in elongated narrow inflorescence. The pods are flattened and usually hairy. They are found on open slopes and meadows up into the alpine zone. Latin name of *Lupinus* comes *lupus*, meaning wolf, alluding to the belief that this plant wolfed nutrients and caused poor soil conditions. To the contrary, lupines are nitrogen fixers that greatly improve soil conditions.

"Quick Key" to the Lupines

1. Stems woody at the base - **2**
1. Stem herbaceous - **3**

2. Petioles 1½ to 4 inches long; raceme 2 to 10 inches long - *L. excubitus*
2. Petioles 3/4 to 1½ inches long; raceme 3 to 12 inches long - *L. albifrons*

3. Plant perennial, with a woody root crown - **4**
3. Plant annual, without a woody root crown -**11**

4. Plant prostrate and matted; leaflets ¼-¾ inch long - *L. breweri*
4. Plant erect; leaflets ¾-4 inches long - **5**

5. Leaflets 10-17 - *L. polyphyllus*
5. Leaflets 5 to 9 - **6**

6. Leaflets 3/8-1½ inches wide and 2-4 inches long - *L. latifolius*
6. Leaflets smaller - **7**

7. Herbage green - **8**
7. Herbage silvery silky - **9**

8. Flowers 3/8-½ inch long - *L. andersonii*
8. Flowers ½-5/8-inch long - *L. formosus*

9. Plant of dry slopes - **10**
9. Plant of wet, damp places - *L. confertus*

10. Plant silky shiny, 20-26 inches tall; pods $1\frac{1}{4}$- $1\frac{1}{2}$ inches long - *L. elatus*
10. Plant silky, less shiny; 8-20 inches tall; pods 1-$1\frac{1}{4}$ inches long - *L. adsurgens*

11. Flowers in separate whorls - **12**
11. Flowers not in separate whorls - **13**

12. Flowers 7/16-5/8-inch long; racemes 2 to 8 inches long - *L. densiflorus*
12. Flowers 3/16-5/16-inch long; racemes 2 inches long - *L. bicolor*

13. Leaves mostly basal; leaflets glabrous above - **14**
13. Leaves well distributed on the stem; leaflets hairy on both surfaces - **15**

14. Flowering racemes $\frac{3}{4}$-3 inches long, well surpassing the leaves - *L. flavoculatus*
14. Flowering racemes $\frac{3}{4}$-inch long, barely surpassing the leaves - *L. brevicaulis*

15. Flowers 3/16-$\frac{1}{4}$- inch long; plant 4-16 inches tall - *L. bicolor*
15. Flowers $\frac{1}{4}$-3/8-inch long; plant 2-8 inches tall - *L. concinnus*

Silky Lupine *(L. adsurgens)* This perennial grows on dry slopes below 5,000 feet in San Diego County.

Silver Lupine *(L. albifrons)* This shrub is 2 to 5 feet tall and occurs in the northern portions of the Transverse Range at about 5,000 feet in sandy and rocky places. Blooms from March to June.

Anderson's Lupine *(L. andersonii)* This is an erect, herbaceous perennial that grows on dry slopes under pines from 4,000 to 8,500 feet in the San Bernardino Mountains and at Mt. Pinos. Flowers from June to September.

Lupine *(L. bicolor)* This erect, long hairy annual is common in open, sandy, and gravelly areas below 5,000 feet. Flowers from March to May.

Short-stemmed Lupine *(L. brevicaulis)* This is a densely, long hairy, almost silvery annual, that grows on sandy places from 4,000 to 7,700 feet in the desert mountains of east San Bernardino County. Flowers from May to June.

Brewer's Lupine *(L. breweri)* This is a low matted perennial with a woody base grows on dry stony slopes between 4,000 to 11,000 feet in the mountains. Flowers from June to August.

Lupine *(L. concinnus)* This is a densely, long hairy, and much branched annual grows on dry, open, and disturbed areas below 5,000 feet.

Lupine *(L. confertus)* This stout perennial grows in mountain meadows and in moist places from 3,000 to 8,500 feet on Mt. Pinos and the San Bernardino Mountains. Flowers from June to August.

Lupine (*L. densiflorus*) This erect annual is common on dry open slopes below 6,000 feet. Flowers from April to June.

Silky Lupine (*L. elatus*) This is erect herbaceous perennial plant that grows 20 to 36 inches tall and has silky, silvery pubescence. It grows on dry slopes among pines from 6,000 to 8,700 feet on Mt. Pinos and in the San Gabriel Mountains. Flowers from June to August.

Interior Bush Lupine (*L. excubitus*) This is a low shrub that grows between 4,000 to 8,700 feet. Blooms from April to June. There are several varieties.

Lupine (*L. flavoculatus*) This low annual plant is found from 2,600 to 7,000 feet in east San Bernardino County on dry slopes. Flowers from April to June.

Summer Lupine (*L. formosus*) This erect perennial grows on dry slopes and under pines from 7,000 to 10,000 feet in the San Jacinto and east San Gabriel Mountains. Flowers from June to August.

Broad-leaved Lupine (*L. latifolius*) This is an erect, herbaceous perennial is common in the open woods and is found from Los Angeles County northward below 7,000 feet. Flowers from April to July.

Many-leaved Lupine (*L. polyphyllus*) This stout, erect perennial grows in wet places from 6,500 to 8,500 feet in the San Jacinto and San Bernardino Mountains. Flowers from June to August.

Ecology & Ethnobotany: The pea-like seeds have been wrongly recommended by some authors of

edible plant books as a substitute for peas. Lupines possess many complex alkaloids and should be considered poisonous. Records do indicate, however, that some species have been safely consumed. For example, Weedon (1996) and Scully (1970) indicate that the young leaves and unopened flowers were steamed and eaten with soup by some Native American tribes. But because of hybridization, the edible species can concentrate toxic alkaloids that could result in an unhealthy game of "lupine roulette."

It does, however, appear that some of the alkaloids found in Lupines are removed by cooking and that toxins intensify with age. The toxic principle of Lupines is excreted by the kidneys, and the poisoning is not cumulative. That is, a lethal dose must be eaten at one time to cause death. The poisonous effects produced by Lupines is referred to as lupinosis, with nervousness, labored breathing, convulsions, frothing at the mouth the obvious signs (Muenschner 1940). But until documentation on southern California species is established, Lupines as a food source are not recommended. Many people use the larger, hairy leaved species as an excellent toilet paper substitute.

Warning: Lupines should be considered to be poisonous.

Alfalfa (*Medicago sativa*)

Description: In general, this is described as a hairless, branching perennial or annual herbs with leaves divided into three leaflets. The terminal leaflet evidently longer than the other two. The pods are twisted. Usually found in disturbed areas at the lower elevations.

Ecology & Ethnobotany: Alfalfa can cause bloat in livestock when it constitutes a high percentage of their diet. Saponins found in the leaves may contribute to the problem. Humans should, therefore, use this plant in moderation. The dried and powdered young leaves and flower heads of Alfalfa are nutritious, and can be steeped in hot water to make a bland tea. The tender leaves can also be added to salads and are rich in Vitamins A, D, and K. Alfalfa also supplies calcium, magnesium, and phosphorus. Alfalfa sprouts are a popular salad addition and the seeds are available from various health stores. Nectar from the flowers produces a good honey. In addition to uses as food and medicine, Alfalfa seeds contain an oil for use in paints and varnishes. Paper makers have used the stem fibers in their craft, and wool dyers extract a yellow dye from the seeds. The seeds of another species, *M. lupulina* (Black Medic), can be parched and eaten, or ground into flour.

Sweetclover (*Melilotus*)

Description: Sweetclover are strongly tap-rooted perennial or annual herbs. The leaves are divided into three fine-toothed wedge-shaped leaflets. The white or yellow flowers are loosely arranged in an inflorescence and the pods are thickly spindle-shaped.

Usually found in disturbed habitats at the lower elevations. The genus name is from the Greek *mel*, meaning honey, and *lotus* flower.

Yellow Sweet Clover (*M. indica*) This glabrous plant with yellow flowers is common in waste places in the lower elevations. Flowers April to October.

White Sweet Clover (*M. alba*) This tall, glabrous plant with white flowers is a common weed of waste places, particularly where damp. Flowers from May to September.

Ecology & Ethnobotany: The young leaves (before the flowers appear) of and *M. albus* (Yellow Sweetclover) may be eaten raw or boiled. The fruit may be used as seasoning for soups. The older leaves are toxic and should be avoided. The dried flowering plant of M. officinalis was used in teas for neuralgic headaches, nervous stomach, diarrhea, and aching muscles. Elias and Dykeman (1982) indicate that improperly dried Yellow Sweetclover will easily mold and in the process produce coumarin, an anti-coagulant that can cause severe bleeding and death. Molding Yellow Sweetclover mixed in hay has killed many cattle. Poorly dried or fermented Yellow Sweetclover produces dicoumarol, a potent anticoagulant that is extremely poisonous in excess. It is used in rat poisons.

The plants are sweet scented due to coumarin and become more pleasant when dried. They have been used to scent clothes and protect them from moths as an alternative to moth balls. It has also been a traditional flavoring additive in smoking tobacco and snuff.

False Lupine (*Thermopsis macrophylla*)

Description: This is a stout perennial plant growing 12 to 32 inches tall. The stem and leaves are densely covered with long, white, silky hairs, and the palmately 3-foliate leaves occur on petioles that $1\frac{1}{2}$ inch long. The yellow flowers are in a terminal raceme that is 6 to 10 inches long. Fruit is a pod and is densely covered with short, appressed hairs. False Lupine grows in open places (below 4,500 feet) from Ventura County northward. Flowers from April to June.

Ecology & Ethnobotany: A cold decoction of the leaves was used by Pomo Natives as a wash for sore eyes. A tea made from the leaves, roots, or bark was used by Kashaya women to slow their menstrual flow.

Clover (*Trifolium*)

Description: There are many species of Clover found. In general, they are annual and perennial plants from rhizomes with leaves that are divided into three or more leaflets. The flower colors range from white, pink, yellow, red, or purple, and the seed pods are round to elongated. They are found in various habitats at all elevations. The genus name refers to the three leaflets.

"Quick Key" to the Clovers

1. Flowers not in a head-like cluster - *T. monanthum*
1. Flowers, many, in a head-like cluster – **2**

2. Toothed involucre at base of flowers - *T. wormskioldii*

2. No involucre at base of flowers - *T. longipes*

Clover (*T. longipes*) This erect or spreading perennial grows in wet meadows between 6,000 to 8,000 feet in the San Bernardino and San Jacinto mountains. Flowers June to August.

Clover (*T. monanthum*) This is a glabrous to slightly hairy perennial, with low, matted, or spreading to erect stems, grows in wet places from 5,000 to 11,500 feet from the San Gabriel, San Bernardino, and San Jacinto Mountains northward

Mountain Clover (*T. wormskioldii*) This glabrous perennial grows in wet places below 10,000 feet in the mountains of southern California, but is more common northward. Flowers from May to October.

Ecology & Ethnobotany: All species are nutritious and high in protein, but the flower heads and tender young leaves are hard to digest raw and may cause

bloating. To improve digestibility of the plants, soak them in salt water for several hours or overnight. Leaves prepared this way may be dried and stored for future use. The dried flower heads and seeds can be ground into a flour substitute or extender.

Trifolium was an important food source for many Native Americans. In the spring, explorers and settlers saw them in the meadows picking and eating large quantities of clover. This was an annual event for the Natives, who relished the greens of the spring season. Unfortunately, the non-natives in their ignorance, compared the Native Americans to grazing animals. This is just one of many disparaging comments made concerning misunderstood Native American behavior.

A tonic tea can be made from the dried flowers. Made strong, the tea can be used as a gargle for sore mouths, throat, and a mild sedative. The tea can also be used as a wash for skin ailments. The dried leaves can be smoked.

American Vetch (*Vicia americana*)

Description: Vetch are annual or perennial herbs with trailing to climbing stems. Leaves are pinnately divided with tendrils in place of terminal leaflets. Found in waste places at lower elevations. *Vicia* closely resembles *Lathyrus* (Sweetpea) and requires close examination of the stipules. The stipules of *Vicia* are usually cut into narrow lobes, whereas the stipules of *Lathyrus* are entire to dentate. This trailing, climbing perennial has sparsely pubescent stems, and grows in open places below 5,000 feet. Flowers from April to June.

Ecology & Ethnobotany: Many species contain toxic compounds and therefore should be considered poisonous, however, Kirk (1975) and Craighead et al. (1963) state that the young stems and seeds can be boiled or baked. The seeds of some species contain compounds producing toxic levels of cyanide when digested.

Caution: Because of the poisonous compounds found in Vetch, it is not recommended for eating.

BEECH FAMILY (Fagaceae)

Members of this family are trees and shrubs, either deciduous or evergreen. The flowers are typically unisexual, with the staminate flowers being catkin-like and the pistillate in an involucre. The fruit is a nut fused to a cup-like organ (cupule) that is composed of fused bracts. There are approximately 8 genera and 900 species in this family. The family is a source of lumber, edible fruits, cork, and many ornamental shade trees.

"Quick Key" to the Beech Family

1. Fruit not an acorn, but a spiny bur - *Chrysolepis sempervirens*
1. Fruit an acorn - **2**

2. Leaves deeply lobed - *Quercus kelloggii*
2. Leaves shallowly lobed; toothed or entire - **3**

3. Shrub - **4**
3. Tree -**5**

4. Young twigs minutely pubescent - *Q. palmeri*
4. Young twigs yellow woolly - *Q. turbinella*

5. Leaves with prominent, parallel lateral veins - *Lithocarpus densiflora*
5. Leaves without prominent, parallel lateral veins - **6**

6. Acorn cap densely woolly - *Q. chrysolepis*
6. Acorn cap not densely woolly - **7**

7. Leaves with shallow lobes and spinose tips - *Q. morehus*
7. Leaves entire or spine tooth, but not shallowly lobed - **8**

8. Leaves convex on upper surface - *Q. agrifolia*
8. Leaves plane on upper surface - *Q. wislizenii*

Bush Chinquapin (*Chrysolepis sempervirens*)

Description: This is a low spreading, evergreen shrub with alternate, oblong to lanceolate leaves. The male flowers occur in catkins that have a rather foul odor, and the fruit is a spiny bur that encloses 1 to 3 nuts. Bush Chinquapin is common on dry slopes in the San Jacinto and San Gabriel Mountains between 2,500 to 11,000 feet. Flowers from July to August.

Ecology & Ethnobotany: The seeds can be eaten raw or stored for winter use. They can also be roasted and pounded into flour. The Paiute made a tea from the leaves.

Tanbark Oak (*Lithocarpus densiflora*)

Description: This tree grows up to 130 feet tall and has red-brown branches and densely woolly young twigs. The leaves have prominent lateral, parallel veins. The male catkins have a foul odor, and the acorn forms at the base of the catkins is cylindric and tomentose when young. The acorn cup is shallow with bristly, spreading scales and tomentose insides. Tanbark Oak occurs in Ventura County and northward on wooded slopes below 4,500 feet.

Ecology & Ethnobotany: Where available, the acorns of Tanbark Oak were considered to be the best tasting. It has been estimated that 100 pounds of

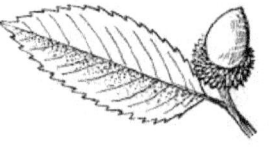 tanbark oak acorns provided about 70 pounds of actual food (meal) that contains about 2,250 calories per pound. For more information on how

to prepare the acorns for food, see the discussion under Oaks (*Quercus*). An infusion of the bark was used as a wash to soothe skin sores or to treat colds and stomach aches.

Oaks (*Quercus*)

Description: There are about 600 species of oaks, with about 60 of them native to the United States, and 19 are native to California. They are described as either trees, shrubs, or shrubs that can be become trees. The leaves are either winter-deciduous, drought deciduous, or evergreen. The margins can be lobed, entire, serrated, or toothed. Fruits are acorns - nuts enclosed by a scaly cup that takes 1 to 2 years to mature.

Coast Live Oak (*Q. agrifolia*) This evergreen tree occurs on rather dry slopes between 2,000 and 4,500 feet in Riverside and San Diego Counties. Catkins appear in March and April.

Canyon Live Oak (*Q. chrysolepis*) This evergreen tree is common on slightly moist slopes and in canyons below 6,500 feet. Catkins appear in April and May.

California Black Oak (*Q. kelloggii*) This deciduous tree with deeply lobed leaves is common in the hills and mountains of southern California between 1,000 to 8,000 feet.

Oracle Oak (*Q. morehus*) This evergreen tree is considered to be a hybrid between *Q. kelloggii* and *Q. wislizenii* and is generally found in the vicinity where they are growing at elevations below 5,000 feet.

Palmer Oak (*Q. palmeri*) This evergreen shrub grows in dry areas from 3,000 to 5,000 feet in the San Bernardino Mountains, the San Gabriel Mountains, and at the west edge of the Colorado Desert.

California Scrub Oak (*Q. turbinella*) This evergreen shrub grows in pinyon-juniper woodlands, between 4,000 to 6,000 feet in the San Bernardino Mountain and southward.

Interior Live Oak (*Q. wislizenii*) This evergreen tree is common on slopes and in valleys below 5,000 feet from Ventura County northward. Flowers March to May.

Ecology & Ethnobotany: With the exception of the highest mountains and the desert areas in the southwestern part of the state, oaks are an abundant tree in California. These plants were particularly important to many Native Californian's as a primary food source. In some places, the acorns comprised as mush as 75% of their daily diet.

Coast Live Oak (*Q. agrifolia*) coast live oaks acorns, which take only 6 to 8 months to mature, were considered the best tasting evergreen oak acorns.

Canyon Live Oak (*Q. chrysolepis*) an evergreen tree that may live over 300 years. Canyon Oaks produce 150 to 200 pounds of acorns a year. Acorns take 18 months to mature.

California Black Oak (*Q. kelloggii*) this winter deciduous species may produce 200 to 300 pounds of acorns a year. These acorns were prized by the Natives as they were good tasting and stored well.

Interior Live Oak (*Q. wizlizenii*) the acorns of this evergreen species take two years to mature, and may produce 200 pounds of acorns a year.

The acorns of oaks were a very important food staple for many Native People's. At least half of the Native American tribes in the United States ate acorns, but they were a staple for those in California. There are definite differences in the flavor of acorns and you may also develop preferences as did the Natives. In fact, even people today when collecting acorns, may pass up nearby oak groves of less favored species and travel further for stands offering a preferred acorn.

The acorns of most species ripen in the fall. In most cases, long poles can be used to knock off acorns from the trees. Once collected, the acorns can be cracked with a stone or teeth and the spread out in the sun to dry. The Natives often shelled the acorns before carrying them home so as to lighten the load. However, other groups did not.

Once acorns were brought back to camp they were stored in granaries (e.g., platforms, bare rock or ground, woven baskets, etc.) in such a way as to prevent molding. In more arid climates, basket-like granaries might be less tightly constructed. Where insects were a problem, the leaves of such plant as California Bay (*Umbellularia californica*) or wormwood (*Artemisia ludoviciana*) were used in the granaries to deter insects.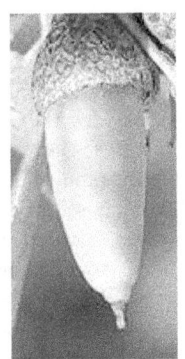

Once the acorns are shelled, the kernels should be rubbed by hand on a basket tray to remove the thin, papery membrane covering it. These were then winnowed, allowing the wind to blow away the light-weight skin while the kernels fall back into the basket, ready to be pounded into flour.

The dry acorns can be pounded using stone pestle (a large flat stone) and a bottomless basket. Another method to produce meal is to use a pestle in conjunction with a mortar (depressions in another rock). Here, the depression is filled with acorn and then

pounded. It is important to try and limit the pounding of the rock as it will then, in the form of grit, become part of the flour. This may be a tedious and time-consuming process, but the food value of the acorns obviously makes it worthwhile. The pulverized meal is winnowed often during the pounding process, with the larger pieces being returned to the mortar for more work.

The pulverized acorn meal contains tannins and glucosides that need to be leached out of the meal in order to make it edible. One of the usual primitive methods in completing this task is to scoop a depression in sandy soil, line the depression with large non-poisonous leaves such as maple, put the meal in the pit and pour water through the meal until the inedible substances are washed away. Another method would be to use loosely woven baskets, perhaps partially buried in sand, to leach the meal.

Sometimes, the acorns can be leached before pounding. Some people buried acorns in swampy places for up to a year (until they turned black), then roasted and ate them without further leaching. Other groups left the shelled acorns to mold in a basket, and then buried them in clean sand in a river bed until they turned black and were ready to eat. The boiled whole or broken kernels can also be boiled in water for 45 minutes or so, changing the water several times

After leaching and tasting the acorn meal to be sure it was sweet, the meal is the ready for use. A porridge can be made by putting some meal into a water-tight basket with some water. Then, using rocks that have been heated in a fire and cleaned by dipping them into water, were then dropped into the basket containing the meal. The rocks should be stirred constantly to prevent them from burning the basket, and when cooled, replaced with other hot rocks. This is repeated until the acorn meal takes on the consistency of porridge. Adding more water would result in a soupy concoction.

This porridge or soup can be eaten as is, or additional flavoring such as mushrooms, roots, berries, greens, meat, or dried fish can be added. Acorn meal, before cooking, is approximately 18 to 20% fat, 5 to 6% protein, and 62 to 68% carbohydrates, with the remainder being water, ash, and fiber. As compared to corn and wheat, both of which contain 1 to 2% fat, 10% protein, and 75% carbohydrates, acorn meal is much higher in fat and lower in protein and carbohydrates.

Medicinally, oaks were by Native Americans in various ways. For example, some Native Americans discovered the forerunner to modern penicillin-type drugs. The ground acorn meal was allowed to accumulate mold which was then scraped off and used to treat boils, sores, and inflammations. Oak galls were mashed and boiled and used as an emetic.

Q. agrifolia An infusion of acorns from used to stop diarrhea. The red inner bark was placed directly against an aching tooth to cure toothache, and to tighten loose teeth. A decoction made from the bark was used to bathe sores and boils.

Q. kelloggii A decoction from the inner bark used to relieve rheumatic pain. Bark was pulverized and used to treat burns.

Q. wizlesinii A decoction from the inner bark used to relieve rheumatic pain. Bark was pulverized and used to treat burns.

Q. chrysolepis These acorns were sucked on when one had a cough or sore throat. The tannin soothes the throat and acts as a cough drop.

The bark of many oaks provided dye and tannin used to color, soften, and cure buckskin. Oak galls have been used throughout history as a dye source. They were crushed, steeped in water overnight, gently simmered, then thoroughly strained to produce a rich tan color.

Warning: Unprocessed acorns may be toxic if eaten in quantity.

SILK TASSEL FAMILY (Garryaceae)

These are shrubs with 4-angled stems and opposite, evergreen leaves. The flowers occur in catkin-like racemes and the fruit is a berry. There is one genus (*Garrya*) with perhaps 18 species in North America and the West Indies. Some species are cultivated as ornamentals.

Silk Tassel (*Garrya*)

Description: The genus consists of 14 New World species ranging from the Pacific Northwest to Panama. *Garrya* is a highland genus occurring in chaparral and coniferous forests above the lowland deserts. First discovered by David Douglas in the Pacific Northwest in 1826 and named in honor of Nicholas Garry, the 1st secretary of the Hudson's Bay Company.

"Quick Key" to the Silk Tassel

1. Leaves silky pubescent beneath (hairs straight or wavy) - *G. flavescens*
1. Leaves grayish white woolly beneath (hairs curled) - *G. veatcnii*

> **Silktassel Bush (*G. flavescens*)** This is an evergreen shrub that grows on dry slopes between 3,000 to 8,000 feet in the mountains.
>
> **Southern Silktassel (*G. veatcnii*)** This evergreen shrub grows on dry slopes below 7,000 feet and blooms February to April.

> **Ecology & Ethnobotany**: These plants contain quinine and garryine, and were used medicinally by several Native American groups. The inner bark was used to reduce fever. The leaves were boiled and the resulting infusion for stomach problems and diarrhea. The fire hardened wood can be used as a digging stick.

GENTIAN FAMILY (Gentianaceae)

There are 70 genera and approximately 1,100 species of this family worldwide. Thirteen genera are native to the United States. They are mostly annual or perennial herbs with bitter juice. Several species are cultivated as ornamentals.

Green Gentian (*Frasera*)

Description: Members of this genus are perennial herbs with opposite or whorled leaves. Flowers are bell-shaped and are densely aggregated into pyramid shaped panicles. The two species can be found in dry, open areas or meadows up to the subalpine zone.

"Quick Key" to the Green Gentians

1. Flowers in interrupted, whorled clusters; corolla with purple veins - *F. neglecta*
1. Flowers in an open cluster; corolla with black dots - *F. parryi*

Green Gentian (*F. neglecta = Swertia n.*) This erect, glabrous perennial or biennial plant grows 8 to 16 inches tall. The leaves are opposite and have thin, inconspicuous, white margins. Green Gentian grows on dry slopes in the mountains from 4,500 to 8,000 feet. Flowers from May to July.

Green Gentian (*F. parryi = Swertia p.*) This stout, glabrous perennial grows 2 to 4 feet high. The leaves are opposite and have white margins. Green Gentian is fairly common in dry places from 1,500 to 6,000 feet.

Ecology & Ethnobotany: The fleshy root of a related species, *F. speciosa* (Showy Frasera), can be eaten raw, roasted, or boiled. But, because the root is very bitter, we suggest mixing it with salad greens. An infusion of another related species, *F. albicaulis* (Whitestem Elkweed), was used to treat infected sores.

Description: This is a large genus with most species occurring on moist or wet soil. Members are annual, biennial or perennial herbs from fleshy roots or rhizomes. The flowers are 4 or 5 lobed, tubular- or funnel-shaped. The three species can be found from the foothills to alpine meadows. The genus honors King Gentius of Illyria, ruler of an ancient country on the east side of the Adriatic Sea, who is reputed to have discovered medicinal virtues in gentians.

"Quick Key" to the Gentians

1. Corolla 5-lobed, less than 1 inch long; flowers in clusters - *G. amarella*
1. Corolla 4-lobed, 1-2 inches long; flowers solitary and terminal - **2**

2. Corolla lobes with entire margins; lower leaves numerous and crowded - *G. holopetala*
2. Corolla lobes with fringed margins; lower leaves not crowded at base - *G. simplex*

Felwort (*G. amarella = Gentianella a.*) This is a slender, branched annual grows in moist places from 4,500 to 11,000 feet. Flowers from June to September.

Sierra Gentian (*G. holopetala = Gentianopsis h.*) This is an erect annual that grows in wet meadows from 6,000 to 11,000 feet. Flowers from July to September.

Hiker's Gentian (*G. simplex* = *Gentianopsis* s.) This erect annual or biennial grows in damp meadows in the San Bernardino Mountains and north between 4,000 to 9,500 feet. Flowers from July to September.

Ecology & Ethnobotany: Moore (1979) suggests that Gentians are perhaps the best stomach tonics. As a bitter, Gentians excite the flow of gastric juices, thereby promoting an appetite and aiding in digestion. The root or chopped herb is steeped and drunk before a meal. Herbage and roots of most species is bitter. Craighead et al. (1963), in discussing *G. calycosa* (Rainier Pleated Gentian) make mention of medicinal uses of European and Asian Gentians, and that early settlers used them in much the same way as a tonic. Gentians contain some of the most bitter compounds known, against which the bitterness of other substances is scientifically measured.

GERANIACEAE (Geranium Family)

There are 11 genera and 800 species in this family distributed worldwide. *Geranium* and *Erodium* are native to the United States. Members of this family are 5-merous plants (5 petals, 5 or 10 stamens, pistil of 5 parts). The seedpod resembles head and beak of a stork or crane, hence the common name, with the seeds in the short thickened "head" and the style is elongated into the pointed "beak." The family is of no real economic importance, except as a source of ornamentals, primarily from the cultivated *Geranium* (*Pelargonium*), a tropical genus well developed in South Africa.

Red-stem Storksbill (*Erodium cicutarium*)

Description: Red-stemmed Storksbill is a low growing annual with mostly basal, finely dissected, fernlike, pinnately divided leaves. The flowers are small, pink and mature into the distinctive "stork's bill" fruit. This is an introduced plant that is widespread on disturbed sites at low to middle elevations.

Ecology & Ethnobotany: The leaves of Red-stem Storksbill can be eaten raw in salads or cooked as a potherb. They are particularly palatable when picked young, and have a parsley-like taste. We find it nicely compliments an otherwise bland wild salad and provides a good source of Vitamin K. It is uncertain whether other species of Erodium are edible and it is not recommended.

The species has a reputation of being a diuretic, astringent, and anti-inflammatory herb. The entire plant was used in a warm water bath for persons suffering from the pains of rheumatism. Leaves were also used in a hot tea to increase urine flow and to increase perspiration.

Wild Geranium (*Geranium*)

Description: Geranium s are annual or perennial herbs that are hairy. The leaves are mostly basal, and the flowers are showy, with 5 petals and sepals, and 10 stamens. The mature fruits are spirally coiled. They can be found in wet meadows or dry, open forests. Geranium derived from the Greek word *gernion*, meaning "crane."

Geranium (*G. californicum*) This slender perennial occurs in moist places from 7,000 to 9,000 feet in Ventura County and San Bernardino County. Flowers from June to August.

Wild Geranium (*G. richardsonii*) This glabrous or pubescent perennial grows in moist places between 4,000 and 9,000 feet. Flowers from July to August.

Ecology & Ethnobotany: The leaves and flowers of most species can be eaten, but because of their astringent properties and texture, they are not a choice edible. We find that they are best when tossed in with other greens in salads or steamed as potherbs. In any case, the leaves are better treated as a filler to stretch supplies of other better tasting and less abundant greens. The leaves can also be chopped and added to soups, thereby blending flavors making the leaves more acceptable. Leaves do toughen with age, but are still palatable in stews. Geranium leaves are similar looking to

Monkshood (*Aconitum*), so positive identification of the flowerless plants is important. Harvest leaves and roots from plants identified with flowers.

A leaf or root tea of *G. richardsonii*, which is one of the most widespread western species and frequently hybridizes with other species, can be used as a gargle for a sore throat. The root sliced fresh can be used as a first aid for gum or tooth infections when applied directly on the area of pain.

The herbaceous part of a related species not found in southern California, *G. viscosissimum* (Sticky Geranium), was used as astringent and styptic, internally for diarrhea and hemorrhages. The plant is high in tannins, providing astringent remedies important in traditional medicine for the emergency treatment of injuries and diarrhea. A hot poultice of boiled leaves was used for bruises and skin problems. The green crushed leaves can be applied to relieve pain and inflammation.

CURRANT AND GOOSEBERRY FAMILY
(Grossulariaceae)

This family consists of a single genus (*Ribes*) with approximately 150 species. All are shrubs with palmately lobed leaves. Some species are armed with spines. The family is a source of ornamentals and edible fruits. In many old field guides, *Ribes* is sometimes included as a member of the Saxifragaceae (Saxifrage Family).

Currant, Gooseberry (*Ribes*)

Description: The many species of *Ribes* are shrubs. The species that have prickles on the stems and bristles on the fruit are commonly called Gooseberries. Those without prickles on the stem or bristles on the fruit are Currants. Leaves are palmately veined and shallowly or deeply lobed. The five petals are smaller than the sepals and usually narrowed to a claw-like base. Fruit is a berry.

"Quick Key" to the Currants and Gooseberries

1. Branches with spines at the nodes - **2**
1. Branches without spines at the nodes - **5**

2. Flowers with flat, saucer-shaped tube; twigs bristly, prickly - *R. montigenum*
2. Flowers with cylindrical tube; twigs pubescent - **3**

3. Flowers yellowish to white - *R. velutinum*
3. Flowers pink to purplish - **4**

4. Leaves, twigs, and petioles, and glandular - *R. amarum*
4. Leaves, twigs, and petioles, pubescent, but not glandular - *R. roezlii*

5. Flowers pink - **R. nevadensis**
5. Flowers, greenish white - **R. cereum**

Bitter Gooseberry (R. amarum) This erect, deciduous shrub grows in wooded canyons below 5,000 feet from San Diego County northward.

Wax Currant (R. cereum) This erect, fragrant shrub occurs in dry rocky areas between 5,000 to 12,500 feet in the San Jacinto Mountains northward. Flowers June to July.

Mountain Gooseberry (R. montigenum) This straggly shrub may be found on dry, rocky areas in the mountains in Riverside County and northward from 7,000 to 12,500 feet. Flowers June to July.

Mountain Pink Currant (R. nevadense) This deciduous shrub grows in moist places between 3,000 and 10,000 feet, from San Diego County northward. Flowers from May to July.

Sierra Gooseberry (R. roezlii) This stout shrub grows on dry, open slopes between 3,500 and 8,500 feet from San Diego County northward. Flowers May to June.

Plateau Gooseberry (R. velutinum) This stout, rigid branched shrub with yellow to white flowers grows on dry slopes between 2,500 and 8,500 feet from Los Angeles County northward.

Ecology & Ethnobotany: The berries of almost all species of *Ribes* are edible raw, and none are known to be poisonous. However, we have come across some unpalatable species, berries with an unpleasant odor and a taste to match. The berries are high in Vitamin C and one of the richest plant sources for copper. One method of collecting them in bulk, is by shaking the bushes over sheets of plastic or blankets. Those that are too sour or spiny become more palatable if they cooked or dried. In regard to the fruits with bristles, one can also roll the berries on hot coals in a basket until the bristles have been singed off. When dried, the berries are a great trail snack. The dried berries can also be mixed with meat to make pemmican. The berries contain enough natural pectin to make jelly. The seeds also contain large quantities of gamma-linolenic acid and many herbalists use this oil to treat skin conditions, asthma, arthritis, and premenstrual syndrome. The nectar-filled flowers are considered good trail snacks. The wood makes good arrow shafts.

Leaves of currents and gooseberries may be added to herbal tea blends. The leaves should be fresh or thoroughly dried, not wilted as they may be toxic.

MARE'S-TAIL FAMILY (Hippuridaceae)

The consists of a single genus, *Hippuris*. The genus was at one time assigned to the Haloragaceae (Water Milfoil Family), but it is not closely related.

Mare's Tail (*Hippuris vulgaris*)

Description: This is a common plant in the main mountain chains of the Western United States. The plant at first glance resembles an immature Horsetail (*Equisetum*), but they are unrelated. Horsetails reproduce by spores and have stems that can be quickly pulled apart. The flowers of Mare's Tail are small and inconspicuous. The plant is found in the margins of shallow waters from ponds to lakes to streams. It can also be found in marshy and swampy areas, roadsides, and irrigation ditches.

Ecology & Ethnobotany: The whole plant is edible when prepared as a potherb. The plant parts are tender and can be gathered in any stage, even in winter. Ancient herbalists are said to have employed Mare's Tail for internal and external bleeding.

<u>WATERLEAF FAMILY (Hydrophyllaceae)</u>

The 20 genera and 270 species within the Waterleaf Family are distributed worldwide, except for Australia. The western United States appears to be the main center of diversity. Only a few members in the family are cultivated.

Yerba Santa (*Eriodictyon*)

Description: The genus *Eriodictyon* is made up of 9 species of aromatic shrubs that grow in the southwestern United States and Mexico. The common name of Yerba Santa is often used when referring to any one of the several different species that were used medicinally.

Thick-leaf Yerba Santa (*E. crassifolium*) This evergreen shrub with densely gray woolly leaves grows on dry, gravelly, and rocky slopes below 6,000 feet. Flowers from April to June.

Yerba Santa (*E. trichocalyx*) This aromatic, evergreen shrub with glabrous leaves occurs on dry, chaparral slopes and mountains up to 8,000 feet. Flowers from May to August.

Ecology & Ethnobotany: A weak solution of boiled *Eriodictyon* leaves can be drunk as a refreshing tea.

This is one of the most important medicinal plants used by California Natives. The leaves were brewed into a tea to cure stomach aches, colds, coughs, inflammation of the throat, rheumatic pains, paralysis, and fevers. This tea also purified the blood. Used externally, a leaf decoction relieved sores, reduced fevers, and cured paralysis. A poultice of mashed leaves healed sores, wounds, cuts, abrasions, insect bites, sprains, and rashes caused by Poison Oak.

In some places, the leaves served as tobacco for smoking and/or chewing. The leaves were chewed as a thirst quencher.

Note: It is reported that Yerba Santa proved to be so useful in treating bronchitis, it was listed in the Pharmacopoeia of the United States from 1894 to 1905 and 1916 to 1947, and was listed in the National Formulary from 1926 to 1960.

Waterleaf (*Hydrophyllum occidentale*)

Description: This bristly haired perennial grows 4 to 24 inches tall. The leaves are alternate, pinnatifid into 7 to 15 large leaflets with ovate divisions. Flowers occur in a coiled inflorescence, and the calyx lobes are covered with stiff hairs. The corolla is 5 lobed, white to violet in color, 1/4 to 3/8-inch long. Waterleaf grows from the

Tehachapi Mountains northward on dry or moist slopes from 2,500 to 8,000 feet. Flowers from May to July.

Ecology & Ethnobotany: The young shoots, leaves, and flowers of Waterleaf can be eaten raw, or these and the roots may be cooked and eaten. We find them exceptionally good in salads, or when eaten as a trail nibble. They do have a texture that takes some getting used to.

The leaves can be used as a protective dressing for minor wounds, and are slightly astringent. As a poultice, it can be used for insect bites and other minor skin irritations.

Blue Balls (*Nama rothrockii*)

Description: his is a simple or branched perennial with hispid, glandular stems that are 2 to 12 inches tall. The leaves are alternate, lanceolate to oblong, with coarsely wavy-toothed margins and glandular hispid hairs. Flowers are many, in a terminal head-like cluster. The corolla is lavender-purple, and funnel-shaped. Blue Balls grows on dry, sandy flats and benches from 7,000 to 10,000 feet and is found on the north slope of the San Bernardino Mountains. Flowers from July to August.

Ecology & Ethnobotany: The seeds of a related species (*N. demissum*) were dried, pulverized, and boiled with water to make mush or porridge.

Baby Blue Eyes (*Nemophila menziessi*)

Description: This is an annual with slightly watery or succulent stems that are easily broken. The angular or winged stems grow up and among other plants. The lower leaves are pinnately divided into 5 to 7, entire lobes. The upper leaves are entire or 3-lobed, oblong or spatulate in shape and with short, stiff hairs. Leaves are all opposite. Flowers are solitary in the axils of the upper leaves. Between each of the 5, lanceolate, calyx lobes are a small, reflexed bract. The corolla is pale blue, with a whitish center. Baby Blue Eyes is common on shaded damp slopes at lower elevations. Flowers from April to June.

Ecology & Ethnobotany: The root was used to prepare a decoction to cure asthma by some Native groups.

Phacelia (*Phacelia*)

Description: The many species in California include herbaceous annuals, biennials, and perennials with various degrees of hairiness. Flowers are 5-parted, spirally coiled, with stamens extending beyond the corolla.

"Quick Key" to the Phacelias

1. Leaves pinnately divided or compound - **2**
1. Leaves simple - **4**

2. Flowers dark blue - *P. fremontii*
2. Flowers pale blue, lavender, or white - **3**

3. Leaves gray-green, silky pubescent; leaf divisions lanceolate, prominently veined with parallel, lateral veins - *P. imbricata*
3. Leaves green, at least above; leaf divisions oblong or ovate, without prominent, parallel lateral veins - *P. ramosissima*

4. Flowers dark purple - *P. minor*
4. Flowers violet, blue, or white - **5**

5. Leaves coarsely toothed; flowers white to bluish - *P. longipes*
5. Leaves mostly entire; flowers blue to violet - *P. curvipes*

Phacelia (*P. curvipes*) This branched annual is occasionally found on dry slopes between 3,500 and 8,000 feet.

Phacelia (*P. fremontii*) This annual has glandular hairy stems and grows on dry sandy, or clay soil and is quite common in the desert mountains below 7,000 feet. Flowers from March to May.

Phacelia (*P. imbricata*) This perennial plant is fairly common in dry, rocky places below 6,500 feet.

It occurs from the San Gabriel Mountains northward. Flowers from April to June.

Phacelia (*P. longipes*) This much branched annual grows on dry slopes from 3,000 to 7,000 feet from Santa Barbara south to the San Gabriel Mountains. Flowers from April to July.

Wild Canterbury Bell (*P. minor*) This is an erect annual plant that grows in dry disturbed areas below 5,000 feet. Flowers from March to June.

Phacelia (*P. ramosissima*) This is a coarse perennial that is common in shaded canyons of the

chaparral and in the yellow pine forest below 8,000 feet. Flowers from May to August.

Ecology & Ethnobotany: At least one species, *P. ramosissima* (Branching Phacelia) can be cooked and used as greens. However, Strike (1994), suggests that the stems and leaves of *Phacelia* may have been eaten raw, but were most likely cooked. The boiled roots of *P. ramosissima* were used to cure coughs and colds, and to alleviate lethargy. A decoction was also used as an emetic and to relieve stomach aches.

Turricula (*Turricula parryi*)

Description: This is a tall, stout perennial plant, with coarse, glandular, hairy stems that grows 3 to 7 feet high. The leaves are alternate, lanceolate, toothed or entire, and without petioles. The flowers are numerous in a coiled inflorescence. The purple corolla is funnel-shaped, and shallowly 5-lobed. Turricula grows on dry or disturbed places up to 8,000 feet, and is common along roadsides. Flowers from June to August.

Ecology & Ethnobotany: The Kawaiisu Natives used a leaf infusion externally to relieve rheumatic pains and to reduce swellings.

Caution: Beware of picking or touching *Turricula* as it can cause a dermatitis similar, if not worse, to Poison Oak with some people. To many people it is also know as Poodle-dog Bush.

ST. JOHN'S-WORT FAMILY (Hypericaceae)

There are 40 genera and 1,000 species worldwide. The family is of little economic importance in North America. A few species are used as ornamentals.

St. John's-wort (*Hypericum*)

Description: The two perennial species have yellow flowers, and small, translucent glands on the leaves and petals. The species can be found in moist areas at various elevations.

Tinker's Penny (*H. anagalloides*) This straggly, often matted annual or perennial plant grows in wet places in the mountains from 4,000 to 10,000 feet. Flowers from June to August.

St. John's-wort (*H. formosum*) This slender erect perennial grows on wet stream banks and in mountain meadows from 4,000 to 7,500 feet. Flowers from June to August.

Ecology & Ethnobotany: The largest and most widespread of the species is *Hypericum peforatum* (St. John's-wort). Weedon (1996) indicates that the leaves may be eaten fresh or may be dried and ground to a flour that can be used like acorn meal. However, only a small amount of the herbage should be consumed.

Despite its reputation as a weed, St. John's-wort may have much to offer humans as a medicinal plant. A number of clinical studies strongly suggest the plant may be effective in treating depression. Other laboratory studies reveal that St. John's-wort has at least two compounds, hypericin and pseudohypericin, that are active against retro-viruses. As such, it is being closely looked at in Acquired Immune Deficiency Syndrome (AIDS) research. The fresh flowers of St. John's-wort in tea, tincture, or olive oil, were once a popular domestic medicine for the treatment of external ulcers, wounds, sores, cuts, and bruises. The ancient alleged magical properties of St. John's-wort were partly due to the fluorescent red pigment, hypericin, which oozes like blood from the crushed flowers. The red dye and extracts are used in cosmetics.

Caution: Craighead et al. (1963) indicate that white-skinned animals feeding on these plants develop scabby sores and a skin itch. Apparently, the plants contain photosensitive toxins and alkaloids. Therefore, ingestion is not recommended in large quantities.

MINT FAMILY (Lamiaceae)

There are approximately 180 genera and 3,500 species worldwide, with the Mediterranean region being the chief area of diversity. All have epidermal glands that exude an odor when rubbed, not necessarily pleasant and mint-like. The old family name is Labiatae and refers to corolla shape, 2 lips (labia). This family is of considerable economic importance as a source of numerous ornamentals, aromatic oils, and a few weedy genera as well as medicines.

Giant Hyssop (*Agastache urticifolia*)

Description: This is a tall perennial growing 3 to 6 feet tall. The leaves are opposite, ovate in shape, 1 to 3 inches long and 1½ inches wide. They are also coarsely toothed on the margins. Flowers occur in dense whorls and form a terminal spike, 1½ to 6 inches long. The calyx is green or rose, and the corolla is 2 lipped, rose or violet, and 3/8 to 5/8-inch long. Giant Hyssop grows in moist places below 9,000 feet in the San Jacinto Mountain and northward. Flowers from June to August. *Agastache* is Greek (*agan*) meaning much, and *stachys* meaning "ear of grain," referring to the flower cluster.

Ecology & Ethnobotany: The seeds of Giant Hyssop may be eaten raw or cooked, and the leaves can used as a tea or for flavoring stews. The Miwoks of California drank an infusion made from the leaves to relieve rheumatic pain, and for indigestion and stomach pains. The plant is said to have mild sedative qualities. Mashed leaves were made into a poultice for swellings.

Horehound (*Marrubium vulgare*)

Description: This is a woolly perennial herb with bitter sap. The leaves are wrinkled and toothed. The

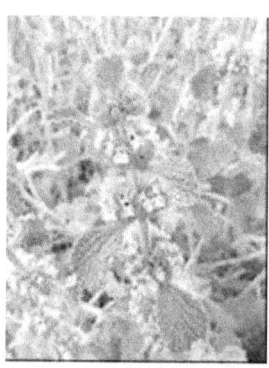

flowers are small, white and occur in dense whorls. Horehound is a common weed of waste places and fields at the lower elevations.

Ecology & Ethnobotany: Horehound is listed as a stimulant, tonic, expectorant, and diuretic. The plant was highly valued by ancient Egyptian priests and the Romans, with the former calling it the "Seed of Horus" and "eye of the star."

The most famous use of this plant is Horehound Candy and is used to soothe sore throats and coughs. A tea from the dried leaves and flowers is also used, but because of the extreme bitterness of the herb, it is obvious why it tastes better in the form of a candy.

Other medicinal uses of the plant include making a warm infusion that will promote perspiration and the flow of urine. When taken cold, this infusion will expel worms.

Mint (Mentha)

Description: All species are distinctly aromatic perennial herbs with rhizomes. The flowers are arranged in whorls. *Mentha* comes from Mintho, the mistress of Pluto, ruler of Hades. Pluto's jealous queen, Proserpine, upon learning of Minthos, trampled her, transforming her into a lowly plant forever to be walked upon. Pluto made this horrible fate more tolerable by willing that the more the plant was trampled, the sweeter it would smell.

Field Mint (M. arvensis) This aromatic perennial with rose to violet flowers is found in moist places below 7,500 feet. Flowers from July to October.

Spearmint (M. spicata) This strongly aromatic perennial with pale lavender flowers is found in moist fields and in marshes below 5,000 feet. Flowers from July to October.

Ecology & Ethnobotany: The fresh or dried leaves of *M. arvensis* (Wild Mint) and *M. spicata* (Spearmint) can be steeped in hot water for a tea. They have also been used as flavoring agents for soups, meat, and pemmican. The young leaves can also be added to salads and soups. The plants are high in Vitamin A, C, K, and minerals iron, calcium and manganese. It is an appetite stimulant and digestive aid. The leaf tea is considered medicinal and was used for colds, stomachaches, fevers, headaches, insomnia, and nervous tension. Crushed leaves were used by some Native Americans to poultice swellings and bruises.

Pure mint oil is a multi-million-dollar industry. It is added to shampoos, massage oil, salves, and soaps, as well as medicines, foods, and liqueurs. It takes approximately 300 pounds of mint to yield 1 pound of oil.

Herb Teas

Plants are the source of practically all the beverages consumed by mankind. Water and milk are the exception. They provide flavor, color, and aroma for endless variety and pleasure. They also provide nutrients for health. Most herbal teas are infusions made by pouring boiling water over herb leaves or flowers and allowing them to steep for 5-10 minutes to release the herb's aromatic oils. A general rule is one teaspoon of dried herb, or 3 teaspoons of fresh crushed herb, per cup of water. To make a stronger tea, add more of the herb rather than steeping the tea longer (long steeping makes the tea bitter). Experiment by combining various herb teas for interesting flavor results.

Monardella (*Monardella*)

Description: These are perennial herbs with leafy stems. Flowers are commonly rose-purple in dense terminal, globose clusters. The bracts are leaf-like, but usually colored.

"Quick Key" to the Monardellas

1. Low plant, no more than 4 inches tall; leaves ovate, toothed $\frac{1}{4}$ - $\frac{1}{2}$ inch long - *M. cinerea*
1. Plant taller than 4 inches high; leaves lanceolate, entire, more than $\frac{1}{2}$ long - **2**

2. Bracts not membranous - *M. viridis*
2. Bracts membranous - **3**

3. Leaves silvery gray - *M. linoides*
3. Leaves green - **4**

4. Plant annual - *M. lanceolata*
4. Plant perennial, woody at the base - **5**

5. Branches 16 to 24 inches tall - *M. linoides*
5. Branches 6 to 16 inches tall - *M. odoratissima*

Gray Monardella (*M. cinerea*) This is a low perennial growing 2 to 5 inches tall. The corolla is pink-rose to purple. Gray Monardella occurs at the highest elevations in the San Gabriel mountains on dry slopes between 6,000 and 10,000 feet. It blooms from July to August.

Mustang Mint (*M. lanceolata*) This erect annual grows 8 to 20 inches tall and has a pleasant mint odor. Mustang Mint grows in dry places in the mountains below 8,000 feet. Flowers from May to August.

Flax-leafed Monardella (*M. linoides*) This perennial plant grows on dry slopes between 3,000 to 9,500 feet. It blooms from June to August.

Mountain Pennyroyal (*M. odoratissima*) This aromatic perennial is found between 4,500 and 9,000 feet in the San Gabriel, San Bernardino, and San Jacinto mountains. Flowers from June to August.

Green Monardella (*M. viridis*) This aromatic perennial grows on dry, rocky slopes between 1,700 and 6,000 feet in the San Gabriel Mountains. Flowers from June to September.

Ecology & Ethnobotany: The leaves and stalks of *M. odoratissima* were eaten, and a thirst-quenching tea was made from the leaves and flower heads. Medicinally, the tea made from the inflorescence of *M. odoratissima* was used for colds and fevers, and relieved digestive upsets and purified the blood. The tea made from the whole plant used in the first stages of a cold when fever is present is said to help relieve elevated temperatures and toxins through sweat. Leaves and flowers rubbed on exposed skin repels mosquitoes and other biting insects.

Common Selfheal (*Prunella vulgaris*)

Description: This perennial grows 4 to 20 inches tall, and has opposite, lanceolate ovate shaped leaves that are 1to 2 inches long. The herbage is glabrous to short pubescent. Flowers are in a dense, terminal spike in the axils of round, membranous, purple tinged bracts. The calyx is purplish and the corolla is 2 lipped, violet, and 3/8 to ¾-inch long. Selfheal is common in moist woods from May to September at elevation. It is also called Hercules' All-heal, because it is supposed that Hercules learned the herb and its virtues from Chiron.

Ecology & Ethnobotany: The entire plant is edible, raw or cooked. However, we found that it is the young and tender plants collected in the early spring that are best. The crushed leaves can be used fresh or dried to make a tea.

Historically, as the common name implies, the plant has been used as a medicine for almost everything. Herbal uses include an astringent, antispasmodic, tonic and styptic. The tea of the dried plant was also used as a gargle for sore throat. Fresh plants can be made into an antiseptic poultice for bruises and scrapes, because of the high tannin content.

Sage (*Salvia*)

Description: There are two species of Salvia in the southern California mountains. Members of this genus can be annual or perennial herbs or shrubs. Look for Sage at the lower elevations.

Gray Ball Sage (*S. dorrii*) This is a low shrub, 1- 3 feet tall that is strongly aromatic and glandular. Gray Ball Sage occurs on dry slopes from 2,500 to 9,000 feet in the mountains of Los Angeles and San Bernardino counties and northward. Blooms from May to July.

Sage (*S. pachyphylla*) This sprawling shrub is strongly aromatic. Sage grows on dry, rocky slopes from 5,000 to 10,000 feet in the San Bernardino mountains and south. It blooms from July to September.

Ecology & Ethnobotany: The seeds of perhaps all species of *Salvia* may be eaten raw, or parched and ground into meal. The seeds can also be soaked in water for a flavorful drink. Leaves of any fragrant sage can be used as a tea or spice for soups and meats. age does contain moderate amounts of Vitamin A and C, and can be added fresh to salads and sandwiches, however, we advise you to do this sparingly.

Tilford (1993) indicates that the above ground parts are antiseptic, astringent, hemostatic, alterative, and tonic, and make a good strong topical disinfectant and cleansing wash for abrasions, contusions, and chafed skin. They are also an effective gargle for sore throat and congested sinuses.

Note: *Salvia* and some species *Artemisia* are often mistaken for being a "sage." Both are aromatic, but are plants in different families. Sage (*Salvia*) is a member of the Mint Family, while Wormwood (*Artemisia vulgaris*) and Sagebrush (*Artemisia tridentata*) are in the Sunflower Family. Mints have opposite leaves, whereas these *Artemisia* species have alternate leaves.

Skullcap (*Scutellaria siphocampyloides*)

Description: This is a glabrous to slightly pubescent annual, growing 4 to 12 inches tall and with linear lanceolate, opposite leaves. The leaves are sessile,

5/8 to 1 inch long, and with toothless margins. Flowers occur in the axils of the upper leaves. The 2 lipped corolla is deep blue, 1 inch or so long, and curved upward. Skullcap grows on gravelly and rocky soil below 7,800 feet in the San Jacinto and Santa Rosa Mountains. Flowers from May to July.

Ecology & Ethnobotany: A related species, *Scutellaria galericulata* (Marsh Skullcap), has nervine related therapeutic properties and has been used as a remedy for general restlessness. It has been used in acute or chronic cases of nervous tension or anxiety. The calming effects are said to be mild but reliable. A strong tea was also made of *S. lateriflora* (Blue Skullcap) and used as a sedative, nerve tonic, and antispasmodic for all types of nervous conditions. All the species contain scutellarin, the primary active compound that has been confirmed to be a sedative and has antispasmodic qualities. Other species may have similar qualities.

Hedge-nettle (*Stachys ajugoides*)

Description: This perennial grows 2 to 4 feet tall and has stem and leaves covered with bristly hairs. Leaves are triangular to oblong in shape, 2 to $3\frac{1}{2}$ inches long, and with toothed margins. The petioles are 1 to $1\frac{1}{2}$ inches long. Flowers occur in interrupted whorls that form a spike, 4 to 8 inches long. The calyx is covered with long, stiff hairs, and the 2 lipped corolla is rose purple or with purple veins and 5/8-inch long. Hedge Nettle is fairly common in moist places in the woods, below 8,000 feet. Flowers from July to August. Hedge-nettle has no stinging hairs, as do the true nettles (*Urtica*), but resembles them before flowering.

Ecology & Ethnobotany: The leaves and flowers are edible, but because of their fuzzy texture and bitter taste, we find them unpleasant. The tubers can also be eaten raw, cooked or pickled and are best if collected in the autumn. Other species may be edible, but this has not been confirmed. Strike (1994) states generally that *Stachys* tubers were eaten.

The leaves may be soaked in water for a few minutes and used as a poultice (Kirk 1975). Additionally, an infusion of fresh leaves can be used as a wash for sores and wounds.

Vinegarweed (*Trichostema lanceolatum*)

Description: This is a strong-scented annual with lanceolate to lance-ovate shaped leaves. Flowers are light blue and the lower lip is strongly deflexed. This species occurs in dry fields and open places below 5,000 feet. Flowers August to October. From the Greek *trichos* (hair) and *stemon* (stamen), referring to the long slender stamens, a characteristic of the genus.

Ecology & Ethnobotany: This species has a strong pungent vinegar odor and was best known as a fish poison. The fresh plants would be mashed and thrown into the pools or sluggish streams. The intoxicated fish would then float to the surface where they were easily caught by hand. Strike (1994) indicates that the fine hairs on the flowers would catch on the gills of the fish and interfere with respiration, making the fish easier to catch. Medicinally, the leaves were chewed and put in the cavity of an aching tooth. Moore (1979) says that a tea made from the flower tops is good for stomachaches and promotes sweating in dry fevers.

LAUREL FAMILY (Lauraceae)

These are usually aromatic trees or shrubs. The family is largely Tropical and most abundant in southeastern Asia and tropical America. The only representative of the family on the Pacific Coast is California Laurel (*Umbellularia californica*). In the Midwest and Eastern United States, there are Sassafras (*Sassafras*) and Wild Allspice (*Lindera*). The avocado (*Persea*) is also a member of this family. It is native to tropical America but can be found growing from Florida to California.

California Laurel (*Umbellularia californica*)

Description: This is a slender tree growing up to 100 feet tall. The leaves are glabrous, alternate, entire, and oblong-lanceolate. The leaves have a strong pungent odor when crushed. The flowers are small, yellow-green and occur in 6- to 10-flowered clusters. There are no petals and the 6 sepals are green. Tiny orange glands may be observed inside the flower at the base. California Laurel is a fairly common tree at the lower elevations in canyons and more shaded areas below 5,000 feet. Flowers from December to May.

Ecology & Ethnobotany: The flesh of the fruit and the ripe kernel may be eaten. Many Native Americans gathered the ripe fruits and dried them in the sun until the thick outer covering loosened and split. The flesh was eaten as it does not store well. The kernels are rich in oil and can be stored. When needed, they can be roasted and eaten or pounded and then formed into cakes. These cakes can then be dried in the sun and stored for future use. California Laurel leaves may be used for flavoring cooking, but are more potent than the commercially available European Bay (*Laurus nobilis*).

California Laurel leaves contain acrid oils that were used by Natives for various medicinal treatments. The oil was sometimes pressed from the leaves and used to relieve toothaches, earaches, or headaches. A tea from the leaves was used to treat colds, whereas a leaf poultice was used on sores and boils.

To repel insects, the fresh leaves or a leafy branch was used. The boiled leaves were used as a shampoo.

Warning: This species contains volatile oils in the leaves that may cause severe headaches or unconsciousness when inhaled. The oil may also cause skin irritation.

LENNOA FAMILY (Lennoaceae)

This is a small family of non-photosynthetic, succulent, root parasitic herbs. The leaves are short and scale-like. The family consists of three genera with 4 species restricted to California, Arizona, Mexico, and Columbia. The plants occur in sandy soils along the seacoast and in the deserts. In the United States, the family is represented by *Pholisma* and *Ammobroma*. Both plants are small herbs rising for a few inches above the sandy soil, with orange or brown stems and minute violet to purple flowers.

Desert Pholisma (*Pholisma arenarium*)

Description: A rather uncommon parasitic plant on *Croton*, *Eriodictyon*, and some members of the Sunflower Family (Asteraceae), in sandy soils and chaparral up to about 6,000 feet. The genus name is from the Greek, *pholis*, for scale, because of the scale-like leaves.

Ecology & Ethnobotany: The stems were supposedly gathered from February to March and eaten raw, roasted, or baked by the Kawaiisu.

BLADDERWORT FAMILY (Lentibulariaceae)

There are five genera and 300 species in this family worldwide. Two genera, *Utricularia* and *Pinguicula* are native to the United States, and occur in California. They are described as annual or perennial herbs of moist and aquatic habitats. The insectivorous species in this family, trapping by means of sticky leaves and bladders, are sometimes cultivated as oddities.

Bladderwort (*Utricularia*)

Description: About 250 species of Bladderwort occur in the United States. These are aquatic or bog plants with submersed stems. The leaves are finely dissected and the yellow flowers are strongly two-lipped with a spur at the base. They are found growing in ponds, lakes, and sluggish streams at the low to middle elevations.

Ecology & Ethnobotany: Bladderworts are carnivorous plants that entrap small aquatic animals in

their bladders. The bladders are closed at the narrow end by valve-like doors that have stiff trigger hairs on the outer surface. When set, the bladders have a partial vacuum, and when a passing animal touches the bristles, the doors open, the walls of the bladder immediately expand, and the sudden inrush of water captures the prey. The process has been timed at 1/460 of a second. Enzymes digest the trapped victim.

The edibility and uses of Bladderworts is unknown, but several species in this genus are reputed to have diuretic values and have been used to treat dysentery (Strike 1994, Coon 1974).

FALSE MERMAID FAMILY (Limnanthaceae)

The two genera and 12 species in this family are restricted to North America. They are annual herbs found in wet places.

Meadow Foam (*Limnanthes gracilis*)

Description: The delicate, glabrous annual has alternate leaves that are pinnate compound into ovate-lanceolate lobes, and these then further divided into 3 to 5 lanceolate segments. The flowers are, becoming pink in age. Meadow Foam grows on lake shores and in wet places in the mountains of San Diego County from 4,500 to 5,000 feet. Flowers from April to May.

Ecology & Ethnobotany: The oil contained in Limnanthes seeds is like the seed oil from Jojoba (*Simmondsia chinensis*), which is similar to sperm whale oil. The seeds of a related species (*L. alba*) were eaten by Natives Americans.

FLAX FAMILY (Linaceae)

There are about 12 genera and 300 species in this family worldwide. The family is of some economic importance because of flax fibers, linseed oil, and the ornamentals obtained.

Flax (*Linum lewisii*)

Description: This much branched annual has pink to white flowers and alternate, sessile, and linear leaves. This species grows on open slopes from 1,000 to 9,500 feet. Flowers from May to July.

Ecology & Ethnobotany: Flax has had value through the ages for its many uses, such as for thread, fabric, oil, paper money, and cigarette paper.

The seeds contain a cyanide compound, but are edible after roasting them. They have a high oil content that contains essential fatty acids that are very much needed in our daily lives, plus they add an agreeable flavor to cooked foods. The crushed seeds have been used as a poultice for irritation, boils, and pain. An infusion of stems is said to relieve stomach aches or intestinal disorders. The roots were also steeped to make an

eye medicine. The stems are a source of linen, a fabric used for clothing.

MENTZELIA FAMILY (Loasaceae)

There are about 15 genera and 250 species of this family occurring chiefly in South America and the warmer parts of the North America. Various species of Mentzelia are endemic in the western United States.

Blazing Star (*Mentzelia*)

Description: Members of this family have alternate, entire, or pinnately lobed leaves. The fruit is a capsule that opens at the top. There are many species of *Mentzelia* in the western United States. It is also called Stick-leaf because of the barbed hairs on the leaves which readily cling to fabric.

"Quick Key" to the Blazing Stars

1. Petals 2-3 inches long - *M. laevicaulis*
1. Petals less than 1 inch long - **2**

2. Petals emarginate; flowers not hidden by bracts; stems white shiny - *M. albicaulis*
2. Petals not emarginate; flowers hidden by bracts - *M. congesta*

Whitestem Blazing Star (*M albicaulis*) This slender annual has white shining stems. It is common on dry sandy soil below 8,000 feet. Flowers from March to July.

Blazing Star (*M. congesta*) This bristly, hairy, branched annual grows in the mountains in dry, gravelly soils between 4,000 and 8,000 feet. Flowers May to July.

Smoothstem Blazing Star (*M. laevicaulis*) This coarse and stout biennial has shining white stems that are rough hairy above. This species grows on dry, stony, gravelly slopes below 8,500 feet. Flowers from June to October.

Ecology & Ethnobotany: *Mentzelia* was considered an important food source in many places of the West. The seeds are edible after being parched and ground into flour. Murphy (1990) describes a type of "gravy" made from the seeds of *M. albicaulis* (Whitestem Blazing Star) and *M. laevicaulis* (Smoothstem Blazing Star);

> "...the red seed is put into a hot frying pan and when the seeds turn a darker red, warm water is added and it is stirred till it thickens."

The Hopi Indians in the southwest parched and ground the small, oily seeds of *M. albicaulis* into a fine, sweet meal and ate it in pinches.

LOOSETRIFE FAMILY (Lythraceae)

Members of this family are herbs, shrubs, or trees. The leaves are opposite or whorled. There are approximately 25 genera and 550 species widely distributed around the world. Seven genera are native to the United States. The family is a source of dyes and ornamentals.

Loosestrife (*Lythrum californicum*)

Description: This erect perennial grows 20 to 72 inches tall and has pale green, glabrous stems. The leaves are alternate, linear to linear-oblong in shape, entire, and 3/8 to $1\frac{1}{2}$ inches long. Flowers are purple, with cylindric base, and solitary in the leaf axils. The 6 petals, each $\frac{1}{4}$-inch long, and 4 to 12 stamens. Loosestrife grows in moist places below 6,000 feet. Flowers from April to October.

Purple Loosestrife is a Eurasian species that has become a nasty wetland weed. It is also cultivated in gardens by some who are unaware of its potential. It is often called the "beautiful killer" because it can take over wetlands and displace native species. It appears to have some efficacy against gnats and flies, and was reported to calm quarrelsome beasts of burden at the plow if placed upon the yoke (Pojar and MacKinnon 1994).

Ecology & Ethnobotany: A tea made from whole flowering plant of Purple Loosestrife, fresh or dried, is a European folk remedy for diarrhea, dysentery, and a gargle for sore throats. It was also used as a cleansing wash for wounds. Experiments have shown that the plant extracts stop bleeding, and kill some bacteria (Foster and Duke 1990). Other species appear to have been used by Native Americans. For example, *L. californicum* was used by the Kawaiisu Indians in California as a medicine and as a dermatologic aid. The method, however, is not reported. Additionally, *L. hyssopifolia* was used by Maidu Indians in California to expedite healing and to reduce inflammation

284

of mucous membranes. It was also used as a shampoo for the hair, but the method is not reported.

MALLOW FAMILY (Malvaceae)

There are some 85 genera and 1,500 species in this family, most of which occur in the tropics. Twenty-seven genera are native to the United States. The distinctive feature of this family is the uniting of the numerous stamen stalks to form a tube around the pistil that resembles a tree trunk, with the anthers and non-fused filaments as the branches and leaves. This "stamen tree" in the center of the flowers is almost a "never fail" characteristic of this family. The family is of moderate economic importance because of cotton fibers derived from the seeds of *Gossypium*, several ornamentals, and a few food plants.

Mallow is from the Greek word meaning soft and may refer to the soft fuzzy leaf characteristic of so many plants in this family, or to the sticky, soothing juice obtained from the roots of some species.

Cheeseweed (*Malva parviflora*)

Description: These plants are distinguished by their distinctive fruit and seeds, rather than their leaves and flowers. They are introduced annual or biennial herbs that are usually found in waste places at the lower elevations.

Ecology & Ethnobotany: Medicinally, the bruised leaves of Cheeseweed can be rubbed on the skin to treat skin irritations. As a headache remedy, leaves or the whole plant can be mashed and placed on the

forehead. Leaf or root tea can be used for angina, coughs, bronchitis, and stomachaches. The fresh or dried leaves were used as a soothing poultice.

The entire plant of another species, *M. neglecta* (Dwarf Mallow), is edible. The young leaves are particularly good in salads or cooked up as a potherb. The plant is, however, very mucilaginous, and it is often used to thicken soup and may take a little getting used to. Eaten in large amounts, however, may cause digestive disorder. The immature fruits (which look like cheese) can also be eaten raw or added to soups.

Alkali Mallow (*Malvella hederacea*)

Description: This spreading perennial has densely, grayish pubescent herbage. The stems are 4 to 16 inches long and the leaves are alternate, round, triangular or kidney shaped. Flowers are axillary, and have 5 cream or pale-yellow petals. Alkali Mallow grows in alkaline or heavy soil, mostly below 6,000 feet. Flowers from May to October.

Ecology & Ethnobotany: While not reported how, this species was used by the Maidu Indians as a laxative (Strike 1994).

Checker Mallow (*Sidalcea malvaeflora*)

Description: This perennial plant grows 6 to 24 inches tall and has pubescent herbage. Leaves are alternate, with the basal ones being heart shaped to kidney shaped. They are also shallowly 7 to 9 palmately lobed and the lobes have toothed margins. The upper leaves are similar, but much more deeply lobed. Flowers occur in a short, few-flowered raceme. The calyx is hairy, and rose pink with white veins. *Sidalcea* grows in the yellow pine forest of the San Bernardino Mountains. Flowers April to August.

Ecology & Ethnobotany: Kirk (1975) says *S. neomexicana* (New Mexico Checker Mallow) is edible after cooking as greens.

OLIVE FAMILY (Oleaceae)

There are approximately 29 genera and 600 species in this family. Members are cosmopolitan in distribution, particularly well represented in temperate and tropical Asia. Five genera occur in the United States, and the family is of considerable economic importance because of olive, timber, and ornamentals. The family is one of the few with flowers having 4 sepals and petals and 2 stamens.

Arizona Ash (*Fraxinus velutina*)

Description: This tree grows to 30 feet tall and has opposite, pinnate compound leaves with 5 to 7 leaflets. The leaflets are ovate to lanceolate, usually glabrous above and without toothed margins. The flowers are small and without petals, and occur in panicles before the leaves appear. The fruit is a samara about an inch long. This species grows along streams and in canyons below 5,000 feet. Blooms in March to April.

Ecology & Ethnobotany: The wood is useful in making bows. In late spring, *Fraxinus* trees are heavily infested with army worms (*Homoncocnemis fortis*) which were gathered by Native Americans, then parched and eaten.

EVENING PRIMROSE FAMILY (Onagraceae)

There are about 20 genera and 650 species worldwide, with a dozen genera native to the United States. The family is of little economic importance, but a few are considered to be ornamentals. However, oil of evening primrose is obtained from this family and is said to be the world's richest source of natural unsaturated fatty acids. The oil is helpful in cases of obesity, mental illness, heart disease, and arthritis and is advertised widely in natural food publications.

Boisduvalia (*Boisduvalia densiflora* = *Epilobium densiflorum*)

Description: This erect annual is covered with long hairs. The leaves are lanceolate, toothed or toothless, and the flowers are subtended by ovate bracts. Petals are rose-purple. This species grows in moist places below 8,500 feet. Flowers from May to August.

Ecology & Ethnobotany: The seeds can be gathered, parched, pulverized and then eaten. They can also be incorporated into pinole. The Maidu rubbed the plant on their head to relieve headaches.

Enchanter's Nightshade (*Circaea alpina*)

Description: This is a low, slender perennial with opposite leaves. Flowers occur in a bractless raceme. There are two sepals that are turned back, and the petals are notched, and white to pink in color. This species grows in deep woods below 8,000 feet in the San Bernardino mountains northward. Flowers in June and August.

Ecology & Ethnobotany: A related species, *C. lutetiana* (Broadleaf Enchanter's Nightshade), was used by the Iroquis as a dermatological aid (on wounds). They also made an infusion as a wash on injured parts.

Clarkia (*Clarkia*)

Description: These are annuals with brittle stems and purple or red, showy flowers. They are usually found on dry slopes at the lower to middle elevations. The genus honors Captain William Clark of the Lewis and Clark Expeditions to the Northwest in 1806.

"Quick Key" to the Clarkias

1. Petals notched or 2-lobed - *C. xantiana*
1. Petals not 2-lobed or only slightly so - **2**

2. Leaves without petioles; petals not clawed - *C. purpurea*
2. Leaves with petioles; petals clawed - **3**

---

3. Petioles 3/8-inch long; sepals distinct; claw broad with lateral lobes at base; capsule glabrous - *C. rhomboidea*
3. Petioles 1/16 to 1 inch; sepals untied and turned to one side; claw slender; capsule long-hairy - *C. unguiculata*

Clarkia (*C. purpurea*) This erect annual has lavender to red-purple petals that are not clawed. It is common below 6,000 feet and grows in open areas. Flowers from April to July.

Clarkia (*C. rhomboidea*) This erect annual with clawed, pinkish-lavender petals grows on dry slopes below 8,000 feet throughout the southern mountains. Flowers from May to July.

Clarkia (*C. unguiculata*) This erect, glabrous annual has clawed, lavender to salmon or dark red-purple flowers. It is quite common throughout the southern mountains below 5,000 feet. Flowers from May to June.

Clarkia (*C. xantiana*) This erect, glabrous annual has lavender to red-purple clawed petals. It grows on dry slopes between 800 to 6,000 feet on the north slopes of the San Gabriel mountains and northward. Flowers in May and June.

Ecology & Ethnobotany: Strike (1994) states that Clarkia seeds were prized by many California Natives and were often used to make pinole. The roots of many Clarkia species were also eaten.

Fireweed, Willowherb (*Epilobium*)

Description: There are many species of Willowherb. The genus includes annual and perennial plants that have willow-like leaves. The flowers are white or lavender in color with petals that are often notched. Fruits are long, narrow pods that open by 4 slits to release the numerous small, densely hairy seeds. The roots and pods are often needed to make positive identification of the many species. The genus name is from the Greek, meaning "on a pod," describing the elongated ovary bearing the other flower parts on its top. The common name refers to the tufts of hairs at the end of the seed, which is similar to that on willow seeds.

Fireweed (*E. angustifolium*) This robust perennial is common in dry areas following a burn. It also grows in moist areas in the mountains below 9,000 feet and north to Alaska. At higher elevations, the plant is shorter with shorter leaves. Flowers from July to September.

Willowherb (*E. brachycarpum*) This erect annual grows in open, dry disturbed places below 7,500 feet. Flowers from June to September.

Willowherb (*E. ciliatum*) This erect perennial grows in moist places below 11,000 feet. Several varieties of the species also occur in the area. Flowers from July to September.

Willowherb (*E. glaberriumum*) This slender, erect perennial grows along streams and in wet areas from 3,000 to 11,500 feet in the San Jacinto Mountains and north. Flowers from July to August.

Willowherb (*E. oregonense*) This slender, erect perennial occurs in wet, boggy places from 5,000 to 11,500 feet in the San Jacinto and San Bernardino Mountains. Flowers from July to August.

Ecology & Ethnobotany: The dozens of species of *Epilobium* are all reported to be edible for people caught in survival situations, but *E. angustifolium* (Fireweed) and *E. latifolium* (Dwarf Fireweed) are the best known and most commonly consumed species. Food, drink, tinder, twine, and medicine are all provided by these abundant herbs. There are many small and "weedy" species found.   In general, they are survivors in landscapes that have been ravaged by man-made and natural forces (e.g., fires, clearcuts). Soil conditions do appear to affect their flavor. Many Native Americans "owned" good patches of Fireweed and these were passed on to subsequent generations. The most distinctive identifying feature of Fireweed is the unique leaf venation. Unlike other plants, the veins do not terminate at the edges of the leaves, but rather join together in loops inside the outer margins.

The young shoots and leaves of Fireweed may be boiled like asparagus, but are better when mixed with other raw greens for a salad. The leaves, green or dry, make a good tea and are useful in settling an upset stomach. Be careful, the leaves are slightly laxative. The unopened flower buds can be used in the same manner as leaves and stems. The young fruits can also be boiled like green beans and are tasty before the seed fibers form. Mature plants tend to become tough and bitter.

The pith of the stems can also be scraped out and eaten as a snack or as a thickener for soups. If consumed in large amounts, Fireweed is a gentle but effective laxative. The plant contains a relatively high content of Vitamin C and beta-carotene.

Raw roots are a popular food of Siberian Eskimos. A poultice made from the roots of Fireweed can be used on skin inflammations, boils, ulcers, and rashes.

The fibrous inner bark can be used as cordage and tinder material. For use in making cordage, I found the fibers brittle. The seeds have cotton-like hairs and are great for fire starting (tinder) and insulation. Many Northwest Indians used the fluffy seed cotton as a wool substitute, mixing it with mountain goat wool or duck feathers. Willow-herb fluff, however, lacks the qualities of a really fine fiber. The flowers can also be rubbed into rawhide to repel water.

Groundsmoke (*Gayophytum*)

Description: These are slender-stemmed annuals with alternate leaves, the lower ones often being opposite. The flowers are small and white. The various species are found on dry slopes, and on the edges of meadows.

Gayophytum (*G. humile*) This slender, much branched annual grows on dry slopes between 3,000 and 11,000 feet. Flowers from June to August.

Pinyon Groundsmoke (*G. racemosum*) This low, slender annual grows on areas that are drying but which have once been wet. It is found in the San Bernardino Mountains from 5,000 to 11,000 feet. Flowers from July to August.

Ecology & Ethnobotany: An infusion of another species, *G. ramosissimum*, was used to soothe irritated skin (Strike 1994).

Evening Primrose (*Oenothera*)

Description: There are many species in this genus which are annual, biennial, and perennial herbs. The flowers are white or yellow, often opening at night. There are 8 stamens, 4 petals, 4 sepals, and the stigma is globe-shaped to deeply four-lobed. The various species can be found in a variety of habitats up to the subalpine zone.

Ecology & Ethnobotany: Most handbooks on edible plants indicate that at least *O. hookeri* (= *O. elata* ssp. *hookeri*) (Hooker's Evening Primrose) and *O. biennis* (Common Evening Primrose), which have edible roots. These were cooked and eaten as a vegetable when young, becoming tough and somewhat spicy or peppery with age. The leaves of *O. biennis* are also edible as cooked greens, but are not exceptional unless mixed with bland greens to make a more acceptable salad. Harrington (1967) suggests that all species would stand a trial as none are known to be poisonous. The various species are known to hybridize easily making identification at times challenging.

We have cooked and eaten the young seed pods of several species and found them to have an acceptable taste. Olsen (1990) also suggests that many species have seeds that are edible after being parched or ground into meal. Strike (1994) states that seeds and leaves of *Oenothera* were eaten by California Natives.

The leaves, stems, and crushed seeds have an astringent quality to them and can be used as a poultice to heal wounds, and for bruises and piles. The seeds are also high in essential oils and have been shown in clinical studies to be effective for heart disease, asthma, arthritis, alcoholism, and other fatty acid problems. The medicinal uses of the oil in these plants are a recent discovery following scientific research in the 1980's that demonstrated their effectiveness for a wide range of intractable complaints. The oil  contains gamma-linoleic acid (GLA), an unsaturated fatty acid, which assists in the production of hormone-like substances. Evening Primrose oil, in the form of gel caps, is becoming popular in the natural supplement's marketplace. Additionally, the stringy bark makes good cordage material.

BROOM-RAPE FAMILY (Orobanchaceae)

There are 13 genera and 180 species found around the world, with four of genera being native to the United States. Members of the Broom-rape Family are herbaceous, lack chlorophyll, and are parasitic on the roots of other flowering plants. The family is of no direct economic importance. The family name comes from the Greek *orogos*, meaning "a clinging plant" and *acho*, "to strangle."

Broom-rape (*Orobanche*)

Description: These species parasitize the roots of other plants. These fleshy annual plants are nearly white to brownish or purplish in color and lack chlorophyll. The leaves are reduced to scales. Broom-rape are usually found in dry soils, associated with such genera as *Artemisia* and *Eriogonum*.

"Quick Key" to the Broom-rapes

1. Main stem 4 to 8 inches tall; calyx 5/8 to $\frac{3}{4}$-inch long with narrow lobes; pedicels 1/8 to 1 inch long - *O. californica*
1. Main stem 3/4 to 2$\frac{1}{2}$ inches tall; calyx 1/4 to 5/16-inch long with short ovate lobes; pedicels 1$\frac{1}{2}$ to 4 inches long - *O. fasciculata*

California Broom-rape (*O. californica*) This glandular, finely hairy herb grows in moist woods from 3,000 to 8,000 feet. Flowers from May to July.

Clustered Broom-rape (*O. fasciculata*) This fleshy and somewhat sticky, pubescent herb is parasitic on plants such as sagebrush and buckwheat, and grows in moist, woodsy areas between 4,000 and 10,500 feet. Flowers from July to August.

Ecology & Ethnobotany: The entire plant of Broom-rape, roots and all, can be eaten raw. Being succulent plants, they answer for food and drink, and are often called Sand Food. We found them to be better tasting when roasted in the hot ashes of a campfire. Strike (1994) also indicates that the roots of *O. californica* (California Broom-rape) and the entire plant of *O. fasciculata* (Clustered Broom-rape) were eaten.

The decocted blanched or powdered seeds are said to ease joint and hip pain. They can also be used as a toothache remedy. Moore (1979) states that the whole plant is astringent and makes an excellent poultice. Broom-rape is also mildly laxative and has sedative properties. The stalks with the white inner portions removed have been used as pipes. *Orobanche uniflora* (One-flowered Broom-rape) was used to treat numerous ailments including bronchial problems, intestinal upset, toothaches, and rheumatic pain. A decoction of *O. fasciculata* was used as a skin wash to kill lice.

WOOD-SORREL FAMILY (Oxalidaceae)

There are seven genera and over 1,000 species distributed worldwide. Only *Oxalis* is native in the United States. In general, they are small plants with leaf blades divided into 3 heart-shaped segments. Flowers are 5-merous, and yellow or purple. The seedpods split explosively, scattering seeds some distance from the plant. The family is of little economic importance.

Wood-sorrel (*Oxalis*)

Description: The genus description is similar to what has been described above. Two species (*O. corniculata* and *O. pes-caprae*) that commonly occur in the lower elevations as weeds in lawns and gardens, are beginning to become established in some developed areas in the San Bernardino and San Jacinto Mountains, as well as elsewhere. The genus name is derived from the Greek word *oxys*, meaning "sour."

Ecology & Ethnobotany: The leaves and stems of Oxalis may be eaten raw. To make a tasty dessert, collect and allow a mass of the plants to ferment for a while. The plants also contain a high percent of oxalic acid; therefore, it is recommended that one eat the plants sparingly until accustomed to them. One symptom

of too much oxalate is painful or swollen taste buds. The plants are also high in Vitamin C and were used to remedy scurvy. A drink can be made by steeping the leaves in hot water, followed by chilling, and sweetening it.

Oxalic Acid

The tart, lemony taste of wood-sorrel (Oxalis), cacti (Opuntia), lamb's quarters (Chenopodium), amaranth (Amaranthus), knotweed (Polygonum), dock (Rumex) and other species, is due to the presence of soluble oxalic acid. Without proper preparation these plants when eaten in substantial amounts should be considered toxic. However, when properly prepared, these plants are an excellent food source. The soluble oxalic acid, also known as salt of lemon, is what makes the plants tasty as well as dangerous. The oxalic acid is dangerous because of its solubility and its affinity for calcium. The solubility allows the acid to enter the bloodstream, where it promptly combines with calcium to form non-soluble calcium oxalate. This precipitates in the kidneys where it both plugs the tubules and "burns" all cells in contact with it, potentially leading to renal failure and death. The oxalic acid is readily dissolved in heated water and will combine with calcium as readily in that water as in the bloodstream. Adding bone fragments, egg shells, or some other sources of calcium to cooking water will transform the oxalic acid to non-soluble calcium oxalate in the pot, retaining the full flavor, but rendering harmless the acid. If you have no bone fragments or egg shells, just pour out the first water after boiling for a time and replace with fresh water.

PEONY FAMILY (Peoniaceae)

The family has only one genus with approximately 33 species.

California Peony (*Paeonia californica*)

Description: This spring annual has palmately divided leaves. The flowers are large and showy, and the petals are red-brown. The flowers are often subtended by reduced leaves. This plant occurs in woodsy, shaded areas in the lower elevations. The genus name is from the

Greek for Paeon, the physician of the gods who supposedly used the plant medicinally.

Ecology & Ethnobotany: The leaves are edible when cooked as greens. We have found it best to boil the leaves in several changes of water until the bitterness was removed.

Another method utilized by Native Americans was to pick the young leaves before the blossoms appeared in the spring, boil them, and then place in a cloth sack and weigh the sack down in the river with a stone. By allowing the water to run through the sack overnight removed the bitterness.

Medicinally, an infusion of sliced, oven baked roots was taken for indigestion by the Diegueno.

POPPY FAMILY (Papaveraceae)

Most plants in this family are annual or perennial herbs, and sometimes shrubs. The sap is often milky or colored. There are 26 genera and 200 species in this family distributed in the subtropical and temperate areas of the northern hemisphere, particularly in western North America. Thirteen genera are native to the United States. The family is of little economic importance except for *Papaver somniferum* which yields opium and its many derivatives, including morphine and heroin. A few species are cultivated as ornamentals.

Prickly Poppy (*Argemone munita*)

Description: This prickly stemmed annual or perennial herb has yellow sap, and stands 24 to 60 inches tall. The pale green leaves are lanceolate to ovate and deeply lobed, and are prickly on the veins. The upper leaves clasp the stem. The flowers are large and showy, up to 5 inches across. The petals are white, and the fruit is a lanceolate capsule. Prickly Poppy grows in dry areas below 6,000 feet. Flowers July to August.

Ecology & Ethnobotany: The seeds of *Argemone* have been used in the past as food, but they are so difficult to extract that it hardly seems worth it. Medicinally, the ripe seeds of *A. munita* were roasted, mashed, and applied as a salve on burns and abrasions by some Native Americans. The seeds were also pounded and used as a poultice on open sores and as a hemorrhoid remedy. The juice of *Argemone* has a rubifacient and somewhat caustic effect and was used for burning off warts. A tea made from the plant is an analgesic topically and can be applied to sunburns and abrasions to relieve pain and swelling.

Warning: The plants contain toxic alkaloids.

Bush Poppy (*Dendromecon rigida*)

Description: This is a shrub with entire, lanceolate leaves. The yellow flowers are showy and the sepals fall off early. This is a common species on chaparral slopes below 6,000 feet. Flowers from April to June.

Ecology & Ethnobotany: Apparently the Kawaiisu used the seeds for food. Unfortunately, we are not aware of how they prepared them. Otherwise, when preparing tobacco (*Nicotiana*), the Kawaiisu added a leaf or two of Bush Poppy to enhance the strength of the tobacco.

Golden Ear Drop (*Dicentra chrysantha*)

Description: This is an erect glabrous perennial with several coarse stems. The leaves are basal, and are 2 times pinnate. Flowers are irregular, somewhat heart-shaped in an open inflorescence. The corolla is yellow and the outer petals are swollen at the base. The four petal tips are turned outward. This species is quite common on dry slopes and in burned over areas below 7,000 feet. Flowers from April to September.

Ecology & Ethnobotany: The plants are considered to be poisonous and contain several different alkaloids. These alkaloids are found throughout the plant and can cause trembling, staggering, convulsions, and labored breathing. Large quantities can be fatal. A poultice from *D. cucullaria* (Dutchman's Breeches) was apparently made to treat skin diseases (Foster and Duke 1990).

California Poppy (*Eschscholzia californica*)

Description: This glaucous herb has leaves that are several times dissected into linear segments. The flowers are bright yellow to orange, and showy. This is a common plant on grassy slopes at the lower elevations. Some plants have escaped gardens and can be found growing wild. Flowers March to June.

Ecology & Ethnobotany: The flowers and leaves were eaten by some Native Americans. The foliage was eaten by gathering the plants before the plants bloomed, leaching them in running water, and then cooking them.

Sap from the fresh root is mildly narcotic. It was apparently used by the Cahuilla Indians in California as a sedative for babies. Additionally, a piece of the root was placed in a tooth cavity to stop toothache. A root extract was used as a wash or liniment for headaches and open sores. If taken internally, the root extract caused vomiting.

Warning: All parts of the plant should be considered toxic.

Cream Cups (*Platystemon californicus*)

Description: This low annual with long, white, hairy stems has opposite leaves growing mostly on the lower part of the plant. The flowers are yellow. Cream Cups grows at lower elevations in the southern California mountains. Flowers March to May.

Ecology & Ethnobotany: The leaves can be eaten as greens.

MOCK ORANGE FAMILY (Philadelphaceae)

There are approximately 17 genera and 130 species in this family mostly found in the Northern Hemisphere, from the Himalayas to North America. Nine of the genera are native to the United States. The family is a source of a few ornamentals.

Mock Orange (*Philadelphus microphyllus*)

Description: Mock Orange is a shrub up to ten feet tall with leaves that are opposite and egg-shaped with three distinct veins. The flowers are white and when in full bloom, the flowers scent the air with a delightfully sweet fragrance reminiscent of orange blossoms. The species is found between 7,000 to 8,500 feet in the San Jacinto and Santa Rosa mountains. The genus is named for the Egyptian King Ptolemy Philadelphus.

Ecology & Ethnobotany: The wood of Mock Orange is strong and hard, and does not crack or warp. It is an excellent wood for making bows and arrows. The leaves and flowers foam into lather when bruised and rubbed with hands, and can be used for cleaning the skin. The plant is otherwise considered poisonous.

PLANTAIN FAMILY (Plantanaceae)

Three genera and 270 species of this family are found worldwide. *Plantago* is widespread in the United States. In general, the family is of little economic importance, but several species of *Plantago* are weeds and one (*P. psyllium*) is the source of seeds used to make a commercial laxative.

Plantain (*Plantago*)

Description: These are characterized as short-stemmed annual or perennial herbs with basal leaves. The flowers are greenish or purplish. Many of the species are introduced from Europe and can be found at the lower elevations, particularly in fields and waste places. *Plantago* means "sole of foot" and refers to the sole shaped leaves of plantain that lie close to the ground as though stepped on. Three species are commonly encountered in southern California.

"Quick Key" to the Plantains

1. Leaves broadly ovate; 3 to 6 inches long - *P. major*
1. Leaves narrower - **2**

2. Leaves lanceolate - *P. lanceolata*
2. Leaves linear; herbage silky pubescent - *P. erecta*

California Plantain (*P. erecta*) This low, villous annual is a common plant of dry slopes. Flowers March to May.

English Plantain (*P. lanceolata*) This perennial plant with basal leaves is a common weed in moist places. Blooms from April to August.

Common Plantain (*P. major*) This is a perennial plant with basal leaves. The leaves are ovate and are longitudinally ribbed with conspicuous nerves that merge at the base and apex. It is a common plant of damp places. Blooms from April to September.

Ecology & Ethnobotany: Because of their reputation as being weeds, Plantains are a forgotten edible to many people. In fact, a lot of effort is spent trying to get rid of the plants from gardens. *Plantago major* (Common Plantain) and *P. lanceolata* (Narrowleaf Plantain) were brought over by European settlers for use as potherbs and medicine. The Native Americans called the plants "white man's foot" because they followed the settlers west. The native species of *Plantain* are uncommon in comparison and it is suggested that they only be used when large populations are found.

As a food, the young leaves of Common Plantain and Narrowleaf Plantain were used fresh or cooked. They contain calcium and other minerals. One hundred grams of Plantain is said to furnish as much Vitamin A as a large

 carrot. The older leaves may be too fibrous and bitter for use, but they are usable if one is able to remove the fibers. Seeds are tedious to collect in quantity, but can be ground and used as flour substitute or extender.

The leaves and seeds of many species were used medicinally. The foliage contains tannins and iridoid glycosides, notably aucubin, which stimulates uric acid secretion from the kidneys.

The crushed leaves of Common Plantain provide an astringent juice that can be used to soothe wounds, sores, insect bites, and the rash of Poison Oak. The plantain juice is a traditional treatment for earaches. The seeds contain up to 30% mucilage which swells in the gut to act as a bulk laxative and soothes irritated membranes. Rubbing the leaves on one's skin works as a natural, moderately effective insect repellant.

SYCAMORE FAMILY (Platanaceae)

These are large trees with deciduous bark. There is one genus with about 10 species, mostly located in the temperate and subtropical northern hemisphere. A few of the species are cultivated as ornamentals.

Western Sycamore (*Platanus racemosa*)

Description: This is a deciduous tree, 30 to 75 feet tall with smooth pale bark that easily peels and gives a mottled or camouflaged look. Leaves are large, and are deeply 5-lobed and densely tomentose (hairy like cotton) on both surfaces when young. Flowers are unisexual and minute, borne in spherical heads about a $\frac{1}{4}$-inch in diameter. Heads occur in chains on slender peduncles on more or less a zig-zag axis. Fruit is a dense, spherical head of achenes with tails projecting outward, about 1 inch in diameter. Western Sycamore is common along streams and dry creek beds below 4,000 feet.

Ecology & Ethnobotany: The presence of Sycamore trees is an indication that a stream or underground water source can be found nearby. To ease asthma-like breathing problems, the bark was prepared as a tea and drunk in place of water for about a week. The bark tea was also drunk to aid in childbirth. Sycamore wood is considered to be a good fuel wood.

PHLOX FAMILY (Polemoniaceae)

There are about 18 genera and 320 species in this family, found chiefly in North America and particularly in western United States. The family is a source of a few ornamentals.

Collomia (*Collomia*)

Description: These are annual or perennial herbs with simple or branched stems. There are approximately 15 species in western North America and South America. The flowers are funnel-shaped or tubular with throats that abruptly flare into an expanded limb. The genus name is from the Greek, *kolla*, meaning glue, because of the mucilaginous layer on the seeds of most species.

Large-flowered Collomia (*C. grandiflora*) This erect annual has salmon or almost white colored flowers and grows in dry, open, and wooded areas below 8,000 feet. Flowers from April to July.

Narrow-leaved Collomia (*C. linearis*) This is an erect annual has pink to purplish colored flowers and grows in dry places in the San Bernardino Mountains from 3,000 to 10,500 feet. Flowers from May to August.

**Ecology &
Ethnobotany**: From the
roots of Large-
flowered Collomia an
infusion was made for
high fevers.
Additionally, an
infusion of the leaves
and stalks was taken
for constipation and to
"clean out the system".

Narrow-leaved Collomia was used as a
dermatological aid by the Gosiute. They made a poultice
of the mashed plant and applied it to wounds and bruises.

Eriastrum (*Eriastrum*)

Description: These are annual or perennial
herbs with about 64 species occurring in the western
United States. These are late-blooming plants that
generally occur in dry areas. he genus name is from the
Greek, *erion*, meaning wool, and *astrum*, for star - plants
with star-like flowers.

"Quick Key" to the Eriastrum

1. Leaves pinnatifid into sharp pointed lobes; flowers in
many-flowered terminal clusters - *E. densifolium*
1. Leaves entire; flowers in a few-flowered cluster - *E.
sapphirinum*

Eriastrum (*E. densifolium*) This erect, much
branched perennial plant grows on dry slopes from

4,000 to 8,000 feet elevation. Flowers from May to September.

Eriastrum *(E. sapphirinum)* This is an erect, branched annual, growing in dry places from 4,000 to 8,000 feet. Flowers from June to August.

Ecology & Ethnobotany: A skin lotion was made from the flowers and roots of *E. densifolium* to relieve rashes and venereal sores.

Gilia (*Gilia*)

Description: There are many species of Gilia and they are characterized as annual, biennial, or perennial plants. The leaves are mostly alternate, lobed or dissected with the tips acute. The seeds are sticky when wet. They are found in a variety of habitats at various elevations.

"Quick Key" to the Gilias

1. Leaves not forming a basal rosette - *G. anegelensis*
1. Leaves (at least basal ones), pinnately lobed or dissected - **2**

2. Flowers in terminal head-like cluster - *G. capitata*
2. Flowers not in terminal head-like cluster - **3**

3. Flowers large $\frac{1}{2}$ to 1 inch long - **4**
3. Flowers smaller - **5**

4. Corolla lobes less than 1/8-inch wide; 3/16-inch wide; corolla salverform - **G. leptantha**
4. Corolla lobes 3/16- 5/16-inch wide; 1/4 to 3/8-inch long; corolla long funnel-shaped - **G. splendens**

5. Leaves pinnately lobed into narrow, filiform lobes, less than 1/16-inch wide - **G. ochroleuca**
5. Leaves one or more times pinnately lobed into broader or shorter lobes - **6**

6. Basal leaves 2-3 times pinnately lobed - **7**
6. Basal leaves sinuate toothed or once pinnately lobed - **8**

7. Plant low, no more than 8 inches tall; in mountains north of San Diego County - **G. leptantha**
7. Plant tall 12-48 inches tall; mountains of San Diego - **G. caruifolia**

8. Corolla 2-2$\frac{1}{2}$ times as long as calyx; Mt. Pinos and mountains of Mojave Desert - **G. modocensis**
8. Corolla 1$\frac{1}{2}$ - 2 times as long as calyx; mountains from Los Angeles county south to lower California - **G. diegensis**

Gilia (G. angelensis) This erect annual grows in sandy or gravelly soil below 6,500 feet. Flowers from March to May.

Blue Field Gilia (G. capitata) This is a tall, slender annual is fairly common in open, sandy, or gravelly and well drained slopes and flats below 6,000 feet. Flowers from April to May.

Caraway-leaved Gilia (*G. caruifolia*) This erect annual grows in the mountains of San Diego County from 4,500 to 7,500 feet. Flowers from May to August.

Gilia (*G. diegensis*) This erect annual grows in sandy areas from 2,000 to 7,200 feet from Los Angeles county southward.

Gilia (*G. leptantha*) This erect annual grows in sandy and gravelly spots from 5,000 to 7,700 feet in the San Bernardino Mountains. Flowers from June to August.

Gilia (*G. modocensis*) This branched and spreading annual grows in sandy places from 1,000 to 7,000 feet in the Mt. Pinos region and in the western Mojave Desert. Flowers from April to June

Desert Gilia (*G. ochroleuca*) This delicate annual grows on sandy slopes and plains from 2,500 to 5,000 feet. Flowers from April to June.

Gilia (*G. splendens*) This is an erect, branched annual that is common in openings in the woods and grows from 1,000 to 7,000 feet in the San Jacinto Mountains. Flowers from April to July.

Ecology & Ethnobotany: Strike (1994) indicates that Gilia seeds were eaten by many California Natives.

Scarlet Gilia (*Ipomopsis aggregata*)

Description: Scarlet Gilia is a biennial plant 1-3 feet tall. The tubular or funnel-form flowers are red, orange, pink, or white and showy. The plant is usually found on dry slopes up to the subalpine. The species has also been called Skunk Flower because of a faint skunk-like smell from its glandular foliage.

Ecology & Ethnobotany: The plant was used by Native Americans as a tea to treat colds, to make glue, and to treat blood troubles. In Nevada, the principal use of this plant was for the treatment of venereal diseases. The whole plant was boiled for the purpose and a solution was taken as a tea or used as a wash. The whole plant was also boiled by the Ute Indians in Utah to make a glue. A blue dye can be extracted from the roots.

Prickly Phlox (*Leptodactylon californicum*)

Description: This erect shrub grows to 4 feet tall and has stems that are densely clothed with short, prickly needle-like leaves. The leaves are alternate, fascicled, and palmately parted into 5 to 9 thin needle-like segments. The salverform flowers are about an inch across, showy, bright pink to lavender. Prickly Phlox is found on dry, sandy slopes below 5,000 feet. Flowers January to June.

Ecology & Ethnobotany: *Leptodactylon pungens* (Granite Prickly Gilia) was used as a decoction to bathe swellings, sore eyes, and scorpion stings. This latter species occurs on dry slopes between 4,000 to 9,000 feet on the eastern edge of the mountains from San Diego County to Inyo County.

Linanthus (*Linanthus*)

Description: In general, they are low annuals with opposite leaves that are palmately parted into slender segments or reduced to linear blades. They are found on dry, open slopes.

"Quick Key" to the Linanthus

1. Flowers yellow - *L. aureus*
1. Flowers white, pinkish, or lilac (base of flowers may be yellow or purple) - **2**

2. Flowers salverform with long thin tube - **3**
2. Flowers funnelform with short tubes - **5**

3. Flowers in small clusters; floral tube about 2 times as long as the calyx - *L. brevicaulis*
3. Flowers in dense terminal clusters; floral tube more than 2 times length of calyx - **4**

4. Leaves hispid, ciliate; calyx lobes sharp-pointed; bracts ciliate - *L. ciliatus*
4. Leaves pubescent, not ciliate; calyx lobes not sharp-pointed; bracts not ciliate - *L. androsaceus*

5. Flowers sessile - *L. nuttallii*
5. Flowers with pedicles - *L. liniflorus*

Linanthus *(L. androsaceus)* This delicate annual grows on dry slopes below 5,000 feet. Flowers from April to May.

Desert Gold (*L. aureus*) This slender annual grows in sandy areas below 6,000 feet from Inyo and Ventura Counties south. Flowers from March to June.

Mojave Linanthus *(L. breviculus)* This erect annual grows on dry, open slopes below 7,000 feet. Found in the San Gabriel and San Bernardino Mountains. Flowers from May to August.

Whisker Brush (*L. ciliatus*) This is a stiff, erect annual that grows in dry, open places, mostly below 8,000 feet. Flowers from April to July.

Linanthus *(L. liniflorus)* This erect annual is common below 5,000 feet on dry slopes. Flowers from April to July.

Linanthus *(L. nuttallii)* This is a bushy perennial with a woody base that grows on dry, rocky, or brushy slopes from 4,000 to 12,000 feet in the San Bernardino Mountains and north. Flowers from May to August.

Ecology & Ethnobotany: An infusion was made from *L. ciliatus* by the Maidu and Pomo Indians to treat children's coughs and colds. An unheated decoction was drunk instead of water to purify blood.

Slender Phlox (*Microsteris gracilis*)

Description: This small, branched annual, grows 4 to 8 inches high, and is glandular pubescent above. Leaves are entire, and the lower ones are opposite, while the upper ones are alternate. The leaves are oblong lanceolate. Flowers are small, usually in pairs in the upper leaf axils. The corolla is salverform, with yellow tube and white to purplish pink lobes. *Microsteris* is common in open, grassy areas below 10,000 feet. Flowers from March to August.

Ecology & Ethnobotany: Slender Phlox was eaten by Miwoks in California as greens. Maidu also used Slender Phlox as a poultice on bruises and wounds.

Phlox (*Phlox*)

Description: The plants in this genus are either low shrubs, perennials, or annuals with opposite leaves. The flowers are salver-form in shape. The many species can be found in various habitats at all elevations.

"Quick Key" to the Phlox

1. Leaves 3/8 to ¾-inch long; herbage gray - *P. austromontana*
1. Leaves 1/4 to 3/8-inch long; herbage yellow-green - *P. diffusa*

Phlox (*P. austromontana*) This low, matted or tufted perennial has leaves that are densely crowded and almost appear spiral or whorled. Phlox grows in dry, rocky places in the San Gabriel mountains and southward at 4,500 to 8,000 feet. Flowers from May to July.

Spreading Phlox (*P. diffusa*) This low matted perennial plant has stems 3 to 9 feet long. Spreading Phlox grows on dry slopes and flats from 3,000 to 11,000 feet in the San Gabriel mountains and north.

Ecology & Ethnobotany: A decoction from the roots of a related species, *P. longifolia* (Longleaf Phlox) and other species, were used by some Native Americans as an eyewash for sore eyes. The scraped roots were soaked in water or steeped or boiled to make the wash.

BUCKWHEAT FAMILY (Polygonaceae)

In the Buckwheat Family there are 40 genera and 800 species, of which 15 genera are native to the United States. The family is best represented in the western states. The economic products include food plants and a few ornamentals.

Wild Buckwheat (*Eriogonum*)

Description: These are annual or perennial herbs; some species are woody at the base. The flowers are small and usually bright colored. The many species of *Eriogonum* can be found in various habitats at all elevations. The genus name is from the Greek *erion*, meaning wool, and *gony*, meaning knee or joint, referring to the hairy stems of many species.

"Quick Key" to the Buckwheats

1. Flowers bright yellow - **2**
1. Flowers white, cream, pale yellow, pink, or red - **3**

2. Involucres in umbels; plant of mountains - *E. umbellatum*
2. Involucres one at node; desert mountains - *E. pusillum*

3. Inflorescence forming a ball-like cluster at the tip of the stem - **4**
3. Inflorescence not forming a ball-like cluster - **5**

4. Ball-like cluster 3/4 to 1$\frac{1}{2}$ inches across - *E. ovalifolium*
4. Ball-like cluster less than 3/8 inch across - *E. kennedyi*

5. Involucres in umbels - *E. umbellatum*
5. Involucres not in umbels -**6**

6. Leaves with stiff hairs, not white woolly; flowers and involucres minute, less than 1/16-inch long - *E. parishii*
6. Leaves white or gray woolly, at least below; flowers and involucres more than 1/16-inch long - 7

7. Stems white woolly - **8**
7. Stems glabrous - **10**

8. Stems 2 to 6 feet tall - *E. elongatum*
8. Stems less than 2 feet tall - **9**

9. Leaves round, more than 1/8-inch across - *E. saxatile*
9. Leaves lanceolate, 1/8-inch or less across - *E. wrightii*

10. Leaves lanceolate with acute apex, 1/8-inch or less across - *E. wrightii*

10. Leaves roundish to oblong ovate, $\frac{1}{4}$-$\frac{3}{4}$ inch across; leaf apex rounded, not acute - **11**

11. Annual; leaves round to kidney-shaped with crisped, wavy margins - *E. molestum*
11. Perennial, woody at base; leaves oblong-ovate with slightly wavy-crisped margin - *E. latifolium*

Eriogonum (*E. elongatum*) This perennial which is branched at the base with a white woolly stem is fairly common on dry, rocky areas below 6,000 feet. Flowers from August to November.

Alpine Eriogonum (*E. kennedyi*) This low, matted or tufted perennial, woody at the base and with short, wiry, flowering stems grows on dry, gravelly areas near and above timberline, above 11, 000 on Mt. San Gorgonio, above 10,000 in the San Gabriel Mountains, and 5,000 to 7,000 feet on Mt. Pinos. Flowers from June to August.

Coast Buckwheat (*E. latifolium*) This perennial with a woody base and white flowers grows on dry slopes, 5,000 to 9,000 feet from the Cuyamaca Mountains to the San Bernardino Mountains. Flowers from August to October.

Pine Eriogonum (*E. molestum*) This erect, glabrous annual occasionally occurs in dry places under pines, from 4,500 to 7,000 feet from Ventura County south to San Diego County. Flowers from June to September.

Buckwheat (*E. ovalifolium*) This low, matted perennial with cream, pink, or rose flowers grows on dry, rocky places from 5,000 to 11,500 feet on the desert slopes of the San Bernardino Mountains and northward. Flowers from May to August.

Buckwheat (*E. parishii*) This much branched, delicate annual grows on dry, gravelly slopes from 4,000 to 9,000 feet in the San Gabriel Mountains and southward. Flowers from July to September.

Small Eriogonum (*E. pusillum*) This erect, glabrous annual with bright yellow flowers is common in the desert from 2,000 to 6,500 feet. Flowers from March to July.

Rock Eriogonum (*E. saxatile*) This perennial grows on dry, rocky places from 4,000 to 11,000 feet in the San Jacinto, San Gabriel, and San Bernardino Mountains. Flowers from May to June.

Sulphur Buckwheat (*E. umbellatum*) This perennial with a woody base, basal leaves and a woolly flowering stem grows in the San Bernardino, San Jacinto, Santa Rosa, and San Gabriel Mountains from 3,000 to 9,000 feet. The flowers are cream to bright yellow and red. Several subspecies are known to occur. The species occurs on dry, stony slopes and flowers from July to September.

Eriogonum (*E. wrightii*) This low perennial with branched, woody base and crowded leaves forming a dense mat grows in rocky or gravelly areas at higher elevations, from 5,000 to 10,500 feet, in the San Jacinto Mountains and northward.

Ecology & Ethnobotany: None of the species are known to be poisonous. The flowering stems can be eaten raw or cooked before they have flowered. Seeds can be collected (though tedious) and ground into flour. A tea from the root of *Eriogonum* was used to treat headaches and stomach problems. The plants are mildly astringent and were used as a gargle for sore throats.

Alpine Mountain Sorrel (*Oxyria digyna*)

Description: Alpine Mountain Sorrel is a low perennial with simple roundish leaves clustered at the base of the stem. The flowers are small, red or greenish. The plant is found in cold, wet places among rock crevices in the San Jacinto and San Bernardino mountains between 9,400 to 10,700 feet. Flowers in July to September.

The plant resembles a miniature rhubarb, with small rounded leaves. It has always been highly esteemed in Arctic regions as a "scurvy-grass" with an agreeable sour taste.

Ecology & Ethnobotany: Perhaps one of the most refreshing plants one encounters in the high country is the Alpine Sorrel. The new growth up to flowering time can

be eaten raw, when it tastes like a mild rhubarb. The stems and leaves can be used in salads or prepared as a potherb. Some aboriginal peoples have been known to ferment mountain sorrel as a kind of sauerkraut. This is accomplished by simply letting the plant(s) sit in water for a while. This sauerkraut can then be stored for winter use. The plants were also dried in the sun for traveling. Plants are high in Vitamin C and can be used to prevent and cure scurvy. Large amounts could, however, cause oxalate poisoning.

Knotweed, Smartweed (*Polygonum*)

Description: The many species of Knotweed are annual or perennial herbs with stems that are more or less swollen at the nodes. The flower colors include white, greenish, or pink. They can be found in various habitats up to the higher elevations.

"Quick Key" to the Knotweeds

1. Plant of dry places - *P. douglasii*
1. Plant in water or of moist places - **2**

2. Plant aquatic, in water; leaves 2 to 4 inches long; flowers rose - *P. amphibium*
2. Plant of meadows and along streams; basal leaves 4 to 10 inches long; flowers white - *P. bistoroides*

Water Smartweed (*P. amphibium*) This glabrous perennial with simple, elongate stems grows in ponds and lakes below 10,000 feet. Flowers from July to September.

Knotweed (*P. bistortoides*) This slender, erect perennial with glabrous stems grows in wet places such as wet meadows and along streams from 5,000 to 10,000 feet in the San Jacinto and San Bernardino Mountains and northward. Flowers from June to August.

Polygonum (*P. douglasii*) This slender, erect annual with glabrous stems grows in dry areas from 4,000 to 9,000 feet throughout southern California, and another variety grows in the San Bernardino Mountains between 5,000 to 10,000 feet.

Ecology & Ethnobotany: Experimentation may be the rule for *Polygonum* as none of the species are known to be poisonous. They do, however, vary in degrees of palatability. Tannins are found in the plants and large amounts might cause digestive upset and possible kidney damage. In moderate quantities, however, the genus is generally regarded as safe. Based on our experiments with various species, some have peppery tasting leaves that can be used in flavoring foods. Others have starchy roots that may be eaten raw or boiled, and roasted. Still others have young foliage made into good salads or potherbs. In our opinion, of all the species, *Polygonum bistortoides* (Bistort) tastes the best. This species is very common in mountain meadows.

The seeds have been used whole or ground into flour. The seeds of *Polygonum* are described as a prehistoric food source and are frequently found in archeological remains.

A decoction of the roots can be made for a sore mouth or gums. The root can also be used as an astringent, diuretic, antiseptic and alterative. The roots were eaten by maritime explorers to prevent scurvy. There is a traditional European "Easter pudding" made up of Bistort, nettle, and dock, all of which are high in Vitamin C.

Dock (*Rumex*)

Description: These are annual or perennial herbs. They have small flowers that are greenish and aggregated in a large terminal inflorescence. They can be found in many habitats in the mountains.

Sheep Sorrel (*R. acetosella*) This introduced weed from Eurasia is found between 3,000 to 7,000 feet from the Cuyamaca mountains to the San Gabriel mountains and northward. Flowers March to August.

Curly Dock (*R. crispus*) This weed is usually found at the lower elevations.

Willow Dock (*R. salicifolius*) This species occurs in moist places below 9,000 feet in the southern California mountains and north. Flowers from May to September.

Ecology & Ethnobotany: The young leaves of Dock can be used as greens and we have found that the flavor varies from species to species. The young leaves are best when collected before the flower stalk emerges. Also, because the leaves become watery when cooked, use very little water and don't overcook them. The older leaves in most cases, may be too bitter for use. Euell Gibbons (1964) found that the leaves of Dock are high in Vitamin C and contain more Vitamin A than carrots. Native Americans ground Dock seeds and used the meal to make breads. However, removing the papery seed cover involves a lot of work, and depending on the species, is probably more work than it is worth. The distinctive sour taste of these plants is due to oxalic acid. As with other species that contain oxalic acid, Docks should be used in small portions as they can cause calcium deficiency.

Poisoning from *Rumex* has only been recorded in livestock after large quantities were eaten. Medicinally, the crushed leaves can be applied to boils and the juice of leaves used to treat ringworms and other skin parasites. The juice of the plant and a poultice of the leaves have also been applied to the rash and pain caused by stinging nettles. A poultice of leaves was used for nervous or allergic hives. The fresh roots were boiled in water to provide a decoction for use internally as a laxative. The powdered yellow roots have been used as a tooth cleanser, a laxative, astringent, and an antiseptic (Lewis and Elvin-Lewis 1977). Some *Rumex* roots contain as much as 35% tannin and were used for tanning animal hides.

PURSLANE FAMILY (Portulacaceae)

Nineteen genera and 600 species occur worldwide, of which nine genera are native to the United States. The family is particularly well represented along the Pacific Coast. It is of little economic importance, but includes several ornamentals.

Spring Beauty (*Claytonia lanceolata*)

Description: This glabrous perennial from a globose corm grows up to 6 inches tall. There is one main pair of leaves (opposite), and lanceolate to ovate in shape. The pink to white flowers occurs in open inflorescence. Spring Beauty grows on dry ridges in the San Gabriel and San Bernardino Mountains. Flowers from May to June. The genus is named for Dr. John Clayton, a botanist and notable plant collector of Colonial days.

Ecology & Ethnobotany: Often called Indian Potato, Wild Potato, or Mountain Potato, the small corms can be eaten raw, boiled, or roasted. For many Native Americans, Spring Beauty was an important "root vegetable." When collecting, keep only the largest corms and replant the others. At first, many find the corms distasteful, as they do take a little getting used to. The

corms are high in starch and, when cooked, taste like potatoes. Boil or bake the corm for 30 minutes. Most species are not plentiful, so be conservative in your endeavor. They can also be dried on strings for long-term storage.

The rosettes can also be eaten raw or cooked and are high in Vitamins A and C. They are better when mixed with other salad plants. The leaves of *C. sibirica* (Siberian Springbeauty) were soaked and applied to the head as a remedy for a headache.

Bitterroot (*Lewisia*)

Description: *Lewisias* are indigenous to the western parts of the United States. They can be found clinging precariously to rocky ledges among boulders, on rock-strewn slopes, damp gravely places, alpine meadows, and in near desert conditions where rainfall is seasonal and unpredictable. There are about 18 species, many are evergreens, but other are bulb-like in that they are below ground for part of the year. Several species have large, showy flowers.

Short-petaled Lewisia (*L. brachycalyx*) This species occurs in wet meadows between 4,500 to 7,500 feet in the Cuyamaca and San Bernardino mountains. Flowers may to June.

Lewisia (*L. nevadensis*) This perennial with a fleshy root and several stems grows on wet banks and in moist meadows from 4,500 to 12,000 feet in the San Bernardino Mountains and on Mt. Pinos. Flowers from May to July.

Pygmy Lewisia (*L. pygmaea*) This species occurs in damp gravel between 5,000 to 8,000 feet in the San Bernardino mountains and northward. Flowers July to September.

Bitterroot (*L. rediviva*) This perennial with a fleshy taproot grows from 6,000 to 9,000 feet on Mt. Pinos and in the San Bernardino Mountains. Flowers in May and June.

Ecology & Ethnobotany: Although all species may be edible, *L. rediviva* is the species that has been used extensively. These plants were an important food item for many Natives. The root is remarkably large and thick for a small plant, and contains nutritious farinaceous matter that is much prized. The roots are dug up in spring before flowering. Once dug, the root is peeled promptly and the small red "heart" (embryo of next year's growth) is removed to reduce the roots bitter flavor. It is then steamed, boiled, or pit-cooked and eaten. The root can also be dried and will keep for a long time. The bitterness of the root varies and cooking is said to improve the flavor. The root boiled to a jellylike consistency will be pink in color. The pounded root was chewed for a sore throat.

Though some still collect it today, Bitterroot is considered a rare plant in many areas. There is little evidence, however, that harvesting by Native Americans has contributed to the rare status. Overgrazing and trampling by range livestock and habitat destruction from agricultural encroachment seem to have been a major impact on *Lewisia* populations. Remember, digging the roots destroys the plant. Programs to maintain and enhance habitat for the plant are recommended.

Miner's Lettuce (*Montia*)

Description: The genus is comprised of slightly succulent annual and perennial herbs. The flowers have two persistent sepals and five white or pinkish petals. Most *Montia* species grow in moist, or seasonally wet areas that are partially to fully shaded.

Toad-lily (*M. chamissoi*) This perennial with creeping or floating stems and erect branches grows in wet places, as in meadows from 4,000 to 11,000 feet. Flowers from June to August.

Miner's Lettuce (*M. perfoliata* = *Claytonia p.*) This glabrous annual has the main 2 leaves fused into a circular disk. Miner's Lettuce is common in moist, shaded areas below 5,000 feet. Flowers from April to June.

Ecology & Ethnobotany: All species of *Montia* have stems and leaves that can be eaten raw or boiled like spinach. The roots are also edible raw or boiled. In California, some Native Americans picked Miner's Lettuce and placed it near the nests of red ants. The ants were allowed to crawl over the leaves and were then shaken off. The residue left on the leaves by the ants had an acerbic flavor.

A tea from the leaves was used as a laxative. A poultice made

from the plant was used for rheumatic pains and to stimulate a poor appetite. *Montia perfoliata* (=*Claytonia perfoliata*) (Miner's Lettuce) is one of the few native plants of the United States that has been cultivated elsewhere. Introduced into Europe, Miner's Lettuce is now cultivated and used for salads and as a potherb.

Purslane (*Portulaca oleracea*)

Description: This is a small succulent annual herb found at the lower elevations. It has been used as a food for more than 2,000 years in India and Persia. In Europe, it is grown as a garden vegetable. The genus name

 may be derived from *portula* meaning "little gate," referring to the lid on the capsule.

Ecology & Ethnobotany: The stems and leaves of Purslane have a tart taste. The entire aboveground part of the plant can be boiled, steamed, fried or pickled. The mucilaginous juice of the stems makes a good thickener for soups. Since the plant tends to hold a lot of dirt and grit, you may want to wash it thoroughly. Besides the good flavor, purslane also provides vitamins A and C, iron and calcium. The tiny black seeds are also nutritious. They can be ground and mixed with other flours.

<u>PRIMROSE FAMILY (Primulaceae)</u>

Worldwide, there are approximately 28 genera and 800 species found in this family. Eleven of the genera are native to the United States, mostly to the eastern part of the country. They are of minor economic importance as a source of ornamentals.

Scarlet Pimpernel (*Anagallis arvensis*)

Description: This is a small annual with opposite leaves that clasp the stem and are oval shaped. The small scarlet colored flowers are on solitary stalks. This is an introduced plant from Europe, and in some places of the United States it is quite common and weedy. Look for it in waste places at the lower elevations.

Ecology & Ethnobotany: The genus name is from the Greek Anagelao, which means to laugh. Dioscorides, physician to the Roman army, was said to give the plant to the men to relieve the depression that accompanies disorders of the liver. Another common name applied to this plant is "Poor Man's Weather Glass" because of its habit of closing before a rain.

Scarlet Pimpernel is said to be an effective diuretic which helps eliminate gravel from the kidney and is used in dyspepsia. As a poultice, it was applied to the skin to relieve the itch and sting of insects.

Shooting Star (*Dodecatheon*)

Description: All leaves are basal and form a loose rosette. The flowers are located at the end of a stalk with narrow, reflexed rose colored petals. The species habitats range from grassland to shrubland, meadows, and riparian habitats up to the alpine zone.

"Quick Key" to the Shooting Stars

1. Basal leaves, including petiole, 2 to 6 inches long - **D. alpinum**
1. Basal leaves, including petiole, 8 to 16 inches long - **D. redolens**

Alpine Shooting Star (D. alpinum) This is a glabrous perennial, with a leafless, flowering stems grows in the San Jacinto Mountains and northward from 4,000 to 11,000 feet. Flowers from May to August.

Shooting Star (D. redolens) This glandular, pubescent perennial, has leafless stems, and grows in moist places from 8,000 to 11,500 feet in the San Jacinto Mountains and north. Flowers in July and August.

Ecology & Ethnobotany: Since none of the species are listed anywhere as poisonous, it is likely that all the species are edible. It is usually the texture that

discourages people from using the plants. We have found that the leaves of many species have a good flavor when eaten raw. Weedon (1996) and Strike (1994) also indicate that the roots and leaves of *D. hendersonii* (Mosquito Bills) are edible after roasting or boiling. Scully (1970) believes that at least five species of Shooting Star in the Rocky Mountains were used by American Indians and that they ate the green leaves and roasted the roots. Thompson and Thompson (1972) provide some additional insight into their preparation of *D. jeffreyi* (Tall Mountain Shooting Star);

"... we tried eating the leaves of *Sierra Shootingstar* raw, but decided that their texture made them unappealing to chew on. At least they do not seem to become bitter, even after the flowers are blooming. When boiling for about 15 minutes and seasoned with butter and salt, they make a satisfactory but bland green vegetable."

Very little information regarding the medicinal uses of Shooting Star could be found. The Native Americans of the Northwest used a leaf tea as a treatment for cold sores.

BUTTERCUP FAMILY (Ranunculaceae)

Worldwide, there are 35-70 genera and 2,000 species in the cooler regions of the Northern Hemisphere. Twenty-one genera are native to the United States. Many plants in this family are poisonous, some are grown as ornamentals, and others provide drugs.

Note: Only a few plants in this family were eaten. Most contain an irritating compound, protoanemonin in the fresh leaves, stems, roots, flowers, and seeds. All must be cooked before eating.

Red Baneberry (*Actaea rubra*)

     Description: Red Baneberry is a perennial herb with fibrous roots. The leaves have long petioles and are 2-3 times divided into sharply toothed, lance-shaped segments. The small flowers are white and borne in a branched, congested, hemispheric inflorescence. The fruits are shiny red or white. Red Baneberry is common in moist, montane forests and riparian areas, usually with some partial shade.

     Ecology & Ethnobotany: The entire plant, especially the berries, are poisonous. The plant is sometimes confused with *Osmorhiza chilensis* (Western Sweetroot) which often shares the same habitat. However, unlike Red Baneberry, Sweetroot has a strong licorice-like odor.

     The roots are considered a laxative and can cause vomiting. The roots were also ground, mixed with grease or tobacco and rubbed on the body to treat rheumatism. Ground seeds mixed with pine pitch were applied as a poultice for neuralgia. *Actaea arguta* is described by Moore (1979) as moderately poisonous when taken internally, with cardiac arrest possible from large doses. The powdered root was mixed with hot water and applied as a counterirritant.

     Warning: If large quantities of this plant are consumed, it may cause a cardiac arrest.

Western Columbine (*Aquilegia formosa*)

     Description: The common name is from the Latin *columbinus*, and refers to the flower's resemblance to a cluster of doves. The genus name stems from the Latin name for eagle. It may also stem from the Latin *aqua*, water, and *legere*, to collect, perhaps referring to the nectar that collects at the tips of the spurs.

Ecology & Ethnobotany: The flowers of Western Columbine are edible and have a sweet taste. They can be added to salads in small amounts. Weedon (1996) indicates that the leaves of this species are also edible, but grow bitter with age.

A tea made from the roots of Western Columbine is said to stop diarrhea and the fresh roots can be mashed and rubbed on aching joints. Aboriginal people used various parts of the plants in medicinal preparations for diarrhea, dizziness, aching joints, and possibly venereal disease. The root, boiled with *Ipomopsis aggregata* (Scarlet Gilia) resulted in a brew that induced vomiting. Ripe seeds can be mashed and rubbed into the hair to discourage lice.

Warning: The seeds can be fatal if eaten and most parts of the Columbine contain cyanogenic glycosides. Any therapeutic use of Columbine is strongly discouraged.

Clematis (*Clematis*)

Description: They are herbaceous perennials with erect stems or woody vines. The leaves are opposite or whorled and simple to pinnately compound. The flowers, lacking petals, are solitary or borne in an open, pyramid-shaped inflorescence. Sepals are petal-like. The various species can be found from brushy slopes above creek bottoms to open areas from the low to high elevations.

Southern California Clematis (*C. lasiantha*) This woody vine climbs over other vegetation and is common in canyons and in moist areas below 6,000 feet. Flowers from March to June.

Western Virgin's Bower (*C. ligusticifolia*) This vine is woody at the base and climbs over other plants. It grows in moist areas below 7,000 feet. Flowers from March to August.

Ecology & Ethnobotany: The genus is essentially comprised of poisonous species. Many references list Western Virgin's Bower as poisonous even though the stems and leaves have been chewed by Native Americans as a remedy for colds and sore throats. The plants have a peppery taste and may cause lightheadedness. Tilford (1993) also indicates that Western Virgin's Bower is diaphorhetic, diuretic, and offers unique vasocontrictory or dilating action that makes it useful in the treatment of migraine headaches. The Thompson Indians used the plant to make a head wash for scabs and eczema, and a mild decoction is drunk as a tonic (Teit 1930). Sweet (1976) states that the white portion of the bark was used for fever, the leaves and bark for shampoo, and a decoction of the leaves was used on horses for sores and cuts. The fibers in the bark was used for snares and carrying nets. The dried stalks were used in fire-by-friction sets and the feathery seed tails for tinder.

Caution: The consumption of Clematis may cause internal bleeding. The entire genus contains strong chemical constituents that can irritate skin and mucous membranes.

Larkspur (*Delphinium*)

Description: These are all perennial herbs with tuberous or fibrous roots and erect stems. The leaves are roundish in outline and deeply lobed or divided. The flowers are showy, blue to partly white, containing 5 petal-like sepals with the uppermost prolonged into a spur. There are also 4 petals, 2 partly enclosed by the upper sepals, the lower 2 often hairy and lobed at the tip. They can be found in various habitats, including meadows, thickets, and open woods from the lower to high elevations.

"Quick Key" to the Larkspurs

1. Flowers violet-purple; leaves divided into broad divisions; plant of wet meadows - *D. glaucum*
1. Flowers blue or white; leaves divided into linear divisions; plant of dry places - *D. parishii*

Mountain Larkspur (*D. glaucum*) This glabrous perennial with coarse hollow stems grows in wet meadows and near streams from 5,000 to 10,500 feet in the San Gabriel and San Bernardino Mountains. Flowers from July to September.

Larkspur (*D. parishii*) The erect perennial grows on dry desert slopes below 7,500 feet. Flowers from April to June.

Ecology & Ethnobotany: Cattle and horses can contract the usually fatal disease of delphinosis from eating delphiniums (Muenscher 1939). Plants should therefore be regarded as poisonous. Strike (1994) indicates that some *Delphinium* roots were dried, pulverized, mixed with water, and used by the Kawaiisu Indians in California as a salve on swollen limbs.

Buttercup (*Ranunculus*)

Description: The many species in California are either perennial or occasionally annual herbs with simple to compound leaves. The flowers are solitary or borne in a small inflorescence. The 5 petals are normally yellow or white and have a nectar gland at the base. They can be found in many different habitats from the lower elevations to the alpine zone. The genus name is from the Latin rana for frog and refers to the wet habitat of some species.

"Quick Key" to the Buttercups

1. Petals more than five - **2**
1. Petals not more than five - **3**

2. Leaves heart-shaped, crenate - **R. cymbalaria**
2. Leaves palmately parted - **R. californicus**

3. Plants aquatic; flowers white - **R. aquatilis**
3. Plants not in water; flowers yellow - **4**

4. Leaves palmately parted - **R. eschoscholzii**
4. Leaves simple - **5**

5. Leaves lanceolate or ovate, and entire - **R. alismaefolius**
5. Leaves heart-shaped, and crenate - **R. cymbalaria**

Buttercup (R. alismaefolius) This glabrous perennial grows on wet banks and in meadows from 4,500 to 6,500 feet in the San Jacinto and San Bernardino Mountains. Flowers from June to July.

Water Buttercup (R. aquatilus) This aquatic perennial with leaves and stems submerged in water grows in ponds or in slow-moving streams below 10,000 feet. Flowers from April to August.

California Buttercup (R. californicus) This erect perennial plant grows in meadows and along streams in the San Bernardino mountains and south between 4,500 to 7,500 feet. Blooms May to July.

Alpine Buttercup (R. cymbalaria) This glabrous or slightly hairy perennial plant has erect flowering stems and is found in muddy places below 10,000 feet in the San Bernardino mountains. Blooms June to August.

Alpine Buttercup (R. eschscholtzii) The glabrous perennial with mostly leafless stems grows

 in the San Bernardino and San Jacinto Mountains from 9,000 to 13,000 feet. Flowers from July to August.

Ecology & Ethnobotany: All species are more or less poisonous when raw. The leaves and stems should be boiled in several changes of water to remove the poisonous compounds (Kingsbury 1964). The volatile toxin is also rendered harmless by drying (Craighead et al. 1963). The seeds can be parched and ground into meal for bread or pinole. The roots can also be boiled and eaten and were an important part of some Native American diets. A yellow dye can be obtained by crushing and washing the flowers.

Fendler's Meadow-rue (*Thalictrum fendleri*)

Description: This erect perennial has leaves that are 2-4 times branched into ultimate leaflets that are shallowly lobed or toothed, and closely resemble the leaves of Columbine (*Aquilegia*). This is a fairly common plant in woods at the lower elevations. Blooms from March to June.

Ecology & Ethnobotany: The dried plant of Fendler's Meadow-rue was rolled into a cigarette and smoked, or sprinkled on a fire, to treat headaches (Pojar and MacKinnon 1994). The young leaves of a related species, *T. occidentale* (Western Meadow-rue), are said to be edible (Willard 1992). A tea was made from the roots as a cure for colds and venereal disease. The roots dried and powdered can be used as a shampoo. Additionally, thalicarpine, a substance used in cancer treatment, has been isolated from *T. pubescens*, *T. revolutum*, and *T. dasycarpum*, species found in eastern North America (Mitchell and Dean 1982).

BUCKTHORN FAMILY (Rhamnaceae)

Of the approximately 60 genera and 900 species found worldwide, ten genera are native to the United States. Economically they are of little importance, but several species have edible fruits and others are used as ornamentals.

Snowbush, Buckbrush (*Ceanothus*)

Description: These are shrubs with leaves that are more or less leathery. One important distinguishing feature is that there are 3 prominent veins originating from near the base of the egg-shaped leaves on some species. The flowers are small, blue or white in color. Look for them on open and dry montane slopes.

"Quick Key" to the Snowbushes/Buckbrushes

1. Leaves opposite - **2**
1. Leaves alternate - **3**

2. Leaf upper surface concave - *C. greggii*
2. Leaf upper surface flat - *C. cuneatus*

3. Branches ending in a sharp spine - **4**
3. Branches not ending in a sharp spine or thorn - **5**

4. Shrub 6-12 feet tall; flower clusters 1½ - 3 inches long - *C. leucodermis*
4. Shrub 2-6 feet tall; flower clusters ½ - 1½ inches long - *C. cordulatus*

5. Leaves toothed - **6**
5. Leaves entire - **7**

6. Leaves wavy, less than 5/8-inch long; San Diego county - *C. foliosus*
6. Leaves not wavy, ½-2 inches long; Los Angeles and Riverside Counties north - *C. oliganthus*

7. Leaves deciduous, thin - *C. integerrimus*
7. Leaves persistent, thick and leathery - *C. palmeri*

      Snowbush (*C. cordulatus*) This spreading shrub grows on dry, open flats and slopes at higher elevations from 3,000 to 9,500 feet. Flowers from May to July. This is a rather easy plant to identify and is recognized by its gray appearance and spiny branchlets.

Buckbrush (*C. cuneatus*) This shrub is common on dry slopes of the chaparral below 6,000 feet. Flowers from March to May.

Wavy-leaf Ceanothus (*C. foliosus*) This evergreen shrub grows on dry slopes below 5,000 feet in the Cuyamaca Mountains of San Diego County. Flowers from March to May.

Desert Ceanothus (*C. greggii*) This erect shrub grows on the dry slopes of desert mountains from 3,500 to 7,500 feet. Flowers from May to June.

California Lilac (*C. integerrimus*) This deciduous shrub grows on dry slopes in the mountains from San Diego County northward from 1,000 to 5,000 feet.

Chaparral Whitethorn (*C. leucodermis*) This shrub occurs on dry slopes, rocky slopes below 6,000 feet. Flowers from April to June.

Hairy Ceanothus (*C. oliganthus*) This shrub with flexible, hairy young branches and twigs grows on dry slopes below 4,500 feet. Flowers from February to April.

Palmer's Ceanothus (*C. palmeri*) This evergreen shrub grows on dry slopes between 3,000 to 6,000 feet in the mountains Riverside, Orange, and San Diego Counties. Flowers from May to June.

Ecology & Ethnobotany: The genus has been long recognized as a substitute for commercial black tea and the leaves and flowers could be used to make tea. The seeds can also be used as food. An infusion of the bark may be used as a tonic. Many species contain saponin which gives the flowers and fruits their soap-like qualities. The flowers when crushed and rubbed in water produce a light lather for purposes of washing oneself. Leaves can also be used as a tobacco substitute. The long, flexible shoots were used in basketry. The red roots yield a red dye.

Coffee Berry (*Rhamnus californicus*)

   Description: This shrub grows up to 15 feet tall and has leaves that are elliptic to oval, toothed or entire, and dull or bright green above, white tomentose beneath. Flowers occur in clusters of 6- to 50-flowers and are greenish in color. Fruit is a berry. Coffee Berry grows in canyons between 4,000 to 7,000 feet in San Bernardino County and northward. Blooms from April to July.
   Ecology & Ethnobotany: A decoction of the leaves of this species was used by some Native Americans to soothe rashes caused by Poison Oak. The inner bark provided a purgative and a laxative. To relieve toothache, the Maidu heated a piece of Coffee Berry root and held it against the aching tooth.
   The Kawaiisu used crushed berries of Coffee Berry as a salve on burns, wounds, and sores to prevent infection. The berries were used to stop hemorrhages, counteract poisons, and have a laxative effect.

There were a number of plants used by Native Americans to produce a laxative or purgative effect. This may be related to the foods they ate and the resulting gastrointestinal problems they created. On the other hand, they may be related to a ceremonial or ritual requirement that required purging the body.

ROSE FAMILY (Rosaceae)

The Rose Family consists of approximately 100 genera and 3,000 species worldwide, with the family being particularly common in Europe, Asia, and North America. About 50 genera occur in the United States. The family is of considerable economic importance because of the edible fruits (e.g., apples, pears, cherries, plums, peaches, apricots, blackberries, raspberries, and strawberries among the important fruits) and many ornamentals.

Chamise, Ribbon Bush (*Adenostoma*)

Description: The genus is comprised of 2 species native to California and Baja California. These are shrubs with evergreen leaves that are clustered, needle-like or linear and alternately arranged. The flowers are small, white, saucer-like and massed in a showy terminal inflorescence.

Chamise (*A. fasciculatum*) This shrub is a common chaparral plant and occurs at lower elevations in the mountains. Blooms in the late spring, May to June.

Ribbon Bush (*A. sparsifolium*) This erect tree with drooping branches, shrubby growth, and slender weal trunk with red-brown shredding bark grows on dry slopes below 6,000 feet. It is common in some areas, absent in others. Flowers from July to August.

Ecology & Ethnobotany: Seeds of *A. sparsifolium* can be harvested in July and August and eaten.

An infusion of Chamise was used to cure skin infections. Twigs were ground into a powder, mixed with animal fat, and used as a salve on sores. Chamise leaves were brewed into tea which cured ulcers, colds, and chest ailments.

Chamise, a prominent plant in the chaparral, burns readily. Natives often tightly bound the branches and used them as a torch. Both species created coals that were useful in cooking and roasting food. Gum from a scale insect on Chamise was used as an adhesive to bind arrow points to shafts, baskets to bed mortars, etc.

Serviceberry (*Amelanchier utahensis*)

Description: This shrub grows 3 to 15 feet tall and has white pubescence on the young growth. The leaves are round to oval and are toothed to the base. The flowers are white and the fruit is a purple-black berry that resembles a blueberry. Serviceberry grows on dry slopes between 5,000 to 7,000 feet from San Diego County to the San Bernardino and San Gabriel Mountains. The genus name is the French Savoy word for the Medlar (*Mespilus germanica*), a species that has similar fruits.

Ecology & Ethnobotany: All species within the genus *Amelanchier* produce edible pomes that ripen in late spring and the summer. They were a considered to be a major food for many Native peoples. In fact, some Native Americans intentionally moved their camps to locations where they could be more easily harvested. The pomes may be eaten raw, cooked, or dried. After drying, the pomes can be pounded into loaves or cakes. These in turn may be eaten after softening a piece in water or placing them in soups or stews. Prepared this way, the pomes could be kept for several years. Additionally, the dried pomes could be incorporated into pemmican.

The liquid from the boiled inner bark was an Native American remedy for treating snowblindness. One drop of strained fluid would be placed in an afflicted eye three times daily. It was also used for eardrops and to stop vaginal bleeding. These applications are probably due to the astringency of the plant's tannic acid content.

The wood can be used for arrows, digging sticks, and other useful items. The berry juice makes a purple dye.

Pemmican

Pemmican is a concentrated food used by many early peoples. It is extremely nourishing and does not spoil. To make pemmican, cut meat into strips and dry it until it completely dries and crumbles. It is then ground as fine as possible, usually by pounding. Melted suet (fat) is then poured over the meat, and salt is added for taste as are fresh berries such as currants and serviceberries. The mixture is then kneaded into a paste and packed into containers. In the "old days" the containers were intestines, but you can use plastic-ware. The finished product can be eaten raw, boiled in stews and soups, or fried like sausages.

Mountain Mahogany (*Cercocarpus*)

Description: There are 13 species growing in the western United States and Mexico. Five occur in California and the following two are common. These are shrubs and small trees with hard wood. The flowers are not showy, but instead are sweet with nectar. The fruit is quite characteristic, and is an achene that ends with a long terminal style that is covered with shiny hairs at maturity. The shrubs glisten in the sun from the mass of silvery fruits, each one a "tailed fruit" as indicated by the generic name.

Mountain Mahogany (*C. betuloides*) This evergreen shrub with obovate leaves toothed near the apex is common on chaparral slopes at lower elevations, generally below 6,000 feet. Flowers from March to May.

Curlleaf Mountain Mahogany (*C. ledifolius*) This shrub or small tree with entire, lanceolate leaves is fairly common on dry mountain slopes between 4,000 to 10,000 feet.

Ecology & Ethnobotany: The common name, Mountain Mahogany, applied to this genus is somewhat misleading. These shrubby trees are not related to true mahogany (*Swietenia*), a valuable cabinet-wood of tropical America. The dark reddish brown, mahogany colored hardwood of *Cercocarpus* may have led to this name. Native Americans used the wood for spears, arrow shafts, and digging sticks. The inner brown bark produced a red-purple dye, as did the roots.

A tea to treat colds was prepared by peeling the bark, scraping out its inner layer, and drying and boiling it. The dried sap was pulverized and applied to the ears to treat an earache. A decoction of the bark and leaves was used for women's gynecological problems.

Caution: The leaves and seeds of Curlleaf Mountain Mahogany contain cyanogenic glycosides and should be considered toxic.

Wood Strawberry (*Fragaria vesca*)

Description: These white-flowered perennial herbs are produced from rootstocks and long runners that root at the nodes. Flowers are white to pinkish, and borne in cymes. The leaves are clustered at the base of the stem and are divided into three egg-shaped, coarsely toothed leaflets. Woodland Strawberry is found in moist, humus-rich, well-drained soils of open forest and forest margins up the subalpine zone (4,500 to 7,500 feet).

Ecology & Ethnobotany: The genus name comes

from the Latin *fraga*, the classical name used for the strawberry fruit and referring to its fragrance. The common name, strawberry, comes from the Anglo-Saxon *streawberige* and refers to the berries "strewing" their runners out over the ground.

Strawberries do not keep well and should be dried for future use if not eaten soon after being picked. Tea made from the green or dried leaves is said to tone up one's appetite. It may also be a nerve tonic, and was used for bladder and kidney ailments, jaundice, scurvy, diarrhea, and stomach-aches. Externally, the leaf tea can also be used as an antiseptic wash for eczema and wounds and as a gargle for sore throat and mouth ulcers. The plants do contain substantial amounts of Vitamins A and C and sulphur, calcium, potassium, and iron. To remove tartar, rub the berries on your teeth and let the juice sit for a few minutes. Afterwards, brush your teeth thoroughly with baking soda and water.

Large-leaf Avens (*Geum macrophyllum*)

Description: This is a perennial herb. Moist places such as meadows 5,500 to 8,500 feet, San Bernardino mountains northward.

Ecology & Ethnobotany: This species was used medicinally by a variety of Native Americans. A decoction of the root was taken for stomach pain and a poultice of chewed or bruised leaves was applied to boils and wounds. To others, it was a panacea, where the leaves were chewed as a universal remedy - "good for everything."

Toyon (*Heteromeles arbutifolia*)

Description: This is a small tree or shrub that can grow up to 15 feet tall. The leaves are evergreen, leathery, and glossy. The white flowers occur in flat-topped clusters and the fruits (pomes) are red or sometimes yellow. Toyon occurs at elevations below 5,000 feet in foothill, chaparral, and woodland habitats.

Ecology & Ethnobotany: The fruit, though rather dry and mealy, are edible.

Oceanspray, Holodiscus (*Holodiscus*)

Description: Three species may be found in the southern California mountain. The egg-shaped leaves are shallowly lobed with toothed margins. The creamy-white colored flowers are small and borne in a diffusely branched inflorescence. Mountain spray is common on rocky slopes and in open forests in the middle elevations.

"Quick Key" to Holodiscus

1. Leaves toothed near apex only; inflorescence 1 - 1½ inches long; petioles 1/16-inch long - *H. microphyllus*
1. Leaves toothed to below the middle; inflorescence 1 to 8 inches long; petioles 1/8 - ¾ inch long - **2**

2. Leaves ovate to roundish, 3/8 - ¾-inch long; petioles 1/8-inch long - *H. boursieri*

2. Leaves elliptic to ovate, longer than wide, 2 - 3½ inches long; petioles ¼ - ¾-inch long; common - *H. discolor*

Ocean Spray (*H. discolor*) This shrub grows at lower elevations in the woods or fairly moist areas below 4,500 feet. Flowers from May to August.

Holodiscus (*H. microphyllus*) This bushy shrub grows on dry, rocky places from 5,000 to 11,000 feet in the eastern Mojave Desert, San Gabriel and San Bernardino Mountains. Flowers from June to August.

Ecology & Ethnobotany: The small, dry, one-seeded fruits of Oceanspray can be eaten raw or cooked. This plant is also used as an astringent, diuretic, tonic and emetic. The stem bark may be decocted and drunk to treat upset stomach, diarrhea, colds, and influenza. As a tonic, it is said to give athletes endurance. Pojar and MacKinnon (1994) also indicate that some Native Americans steeped the brownish fruiting clusters in boiling water to make an infusion that was drunk for diarrhea. The hardwood can be used for digging sticks.

Horkelia (*Horkelia bolanderi*)

Description: This is a perennial with a thick woody root that forms tufted or matted clumps. The leaves are pinnate compound and the flowers are white. This species occurs in damp places 4,000 to 9,500 feet, San Jacinto and San Bernardino mountains northward. Another species, *H. californica*, occurs in the San Bernardino mountains and occurs in open slopes and along streambanks below 5,500 feet.

Ecology & Ethnobotany: The Kashaya drank a tea made from the roots of a related species, *H. californica*, to purify blood.

Ivesia (*Ivesia*)

Description: These are perennials with pinnate leaves that are mostly basal. Flowers are yellow, white or purple. About 22 species occur in western North America. The genus is named for Lt. E. Ives, leader of a Pacific Railway Survey.

Silver-haired Ivesia (*I. argyrocoma*) Dry meadows from 6,500 to 7,500 feet in the San Bernardino mountains. Blooms from June to August.

Tahquitz Ivesia (*I. callida*) Found in rock crevices at about 8,000 feet in the Tahquitz Peak area of the San Jacinto mountains. Blooms from July to August.

Mousetail Ivesia (*I. santalinoides*) Occurs on gravelly slopes and ridges, from 6,500 to 9,000 feet in the San Jacinto mountain to Mt. Pinos and northward. Blooms June to August.

Ecology & Ethnobotany: While we have not been able to find any uses for these two species, a related species, *I. gordonii* (Gordon's Ivesia), was used by the Arapaho. They apparently made an infusion of the root for use as a tonic.

Cinquefoil (*Potentilla*)

Description: The many species of Cinquefoil include perennial, biennial, or annual herbs, and one shrub. The flowers are yellow, white, or, in one case, purple. They can be encountered in various habitat types at all elevations.

"Quick Key" to the Cinquefoils

1. Leaves pinnate compound - **2**
1. Leaves palmate compound - **3**

2. Leaflets 9 to 31; white silky below - *P. anserina*
2. Leaflets 5 to 9; not white silky below - *P. glandulosa*

3. Leaflets ovate - *P. wheeleri*
3. Leaflets lanceolate - *P. gracilis*

Silverweed (*P. anserina*) This low, flat-lying perennial with pinnate compound leaves grows in moist, alkaline areas between 4,000 to 8,000 feet in the San Bernardino mountains and north. Flowers May to October.

Sticky Cinquefoil (*P. glandulosa*) This erect perennial with glandular, often reddish stems grows on dry or moist places up to 8,000 feet. Flowers May to June.

Cinquefoil (*P. gracilis*) This perennial with silky, hairy stems grows in moist areas, in meadows, and along streams. It is found between 2,500 to 9,000 feet. Flowers June to August.

Five Finger (*P. wheeleri*) This flat-lying or spreading perennial plant has silky, hairy stems. Five Finger grows at the edge of open meadows from 6,500 to 11,500 feet in the San Bernardino mountains and north. Flowers from June to August.

Ecology & Ethnobotany: The large fleshy, older roots of *P. anserina* (=*Argentina anserina*) (Silverweed) can be boiled or roasted and added to soups and stews. Prepared this way they are quite tasty and have a nutty or a parsnip-like texture, but more woody. They were a staple among many Native Americans. Today they are seldom harvested, but greatly enjoyed by those who still use them. Silverweed is high in tannins and can be used to tan leather. Other cinquefoils are considered astringent as well. For example, a tea can be made from the leaves of *P. fruticosa* (= *Pentaphylloides floribunda*) (Shrubby Cinquefoil). The whole plant or root of *P. arguta* (Tall Cinquefoil) in a tea or poultice, stops bleeding and has been used on cuts, wounds, and for diarrhea and dysentery. A strong tea may still be useful although no longer used as a mouthwash and gargle for sore throats or tonsil inflammations and helps reduce gum inflammation.

Caution: In ancient times these plants were grown for food and medicine. Although there are no reports of toxic reactions from use of this genus, moderation is still advised.

Wild Cherry, Plum (*Prunus*)

Description: This genus has about 400 species. They are shrubs or trees with simple leaves. Many species have a pair of warty glands present at the tops of the petioles or at the bases of the leaf blades. Flowers are pink to white and rather showy. Fruits are drupes with 1 stone and typically embedded in the fleshy pulp. Seeds and leaves are toxic, as they contain hydrocyanic acid.

"Quick Key" to the Plums

1. Branches ending in a sharp spine - *P. fasciculata*
1. Branches not ending in a sharp spine or thorn - **2**

2. Leaves spinose toothed - *P. ilicifolia*
2. Leaves not spinose tipped - **3**

3. Leaves 3/8-1-inch wide; petioles $\frac{1}{4}$-inch long - *P. emarginata*
3. Leaves $\frac{3}{4}$ - 2 inches wide; petioles 3/8 - $\frac{3}{4}$-inch long - *P. virginiana*

Bitter Cherry (*P. emarginata*) This deciduous shrub or tree grows in canyons and moist slopes below 9,000 feet in the mountains. Flowers from April to May.

Desert Almond (*P. fasciculata*) This deciduous shrub grows on dry slopes from 2,500 to 6,500 feet in the desert mountains. Flowers from March to May.

Hollyleaf Cherry (*P. ilicifolia*) This evergreen shrub or small tree is common on chaparral slopes below 5,000 feet. Flowers from April to May.

Western Choke Cherry (*P. virginiana*) This deciduous shrub or small tree grows in damp places in the woods and mountains below 8,000 feet. Flowers from May to June.

Ecology & Ethnobotany: In general, the fruits of these species are sour or bitter when raw, but after cooking the sourness disappears. Native Americans dried the berries whole or in cakes for use in winter. When needed, the dried fruits were soaked in water and then eaten. Lewis and Clark's Expedition members ate Western Chokecherry fruits when other foods were scarce. It seems that after drying, the fruits lose some of their bitterness, resulting in an almost sweet taste.

To make cakes, the ripe fruits are usually ground up, pits and all, and dried in the sun. When needed, the cakes, or portions thereof, can be soaked in water, mixed with flour and sugar and made into a

sauce or gravy. This sauce was eagerly traded among some Native Americans such as the Navajo, Shoshone, Arapahoe, and Ute's. The only difficulty we've found in preparing cakes in this manner is that the pits do not grind down nicely into a fine material, leaving larger chunks that could have resulted in broken teeth.

Other uses of the berries was their incorporation into pemmican. They can also be used in making jelly, but because Chokecherries are low in natural pectin, it is advisable to add pectin.

The leaves of both species contain toxic amounts of cyanide as do the seeds (pits). Cyanide is highly volatile and the pits can be rendered safe by long-term drying, by boiling in several changes of water, or by dry roasting. Do not eat them in significant amounts even then unless you mix them with larger quantities of other foods. Prunus shoots, peeled and split, were used in basketry. The wood was used for various implements, such as digging sticks, arrows, and arrow fore shafts.

Warning: The leaves, bark, and seeds of all Prunus contain cyanide-producing glycosides. Therefore, eating large quantities of ripe berries with their pits could cause nausea and vomiting. In some instances, it could be fatal. Cooking and drying the seeds appears to dispel most of the glycosides and then, the seeds in dried, mashed choke cherries are not as significant a problem. To be safe, it is best to discard the seeds before eating the fruits.

Waxy Bitterbrush (*Purshia tridentata*)

Description: This fragrant shrub grows up to 8 feet tall. Leaves are deeply 3-cleft into linear lobes, glandular above and tomentose below. The leaf margins are rolled inwards. Flowers are pale yellow to white and the fruit is a pubescent oblong achene. Waxy

Bitterbrush grows on dry slopes and canyons of the desert mountains between 3,000 and 9,000 feet. Flowers from April to June. Named for F. Pursh, author of an early flora for North America.

Ecology & Ethnobotany: The leaves and inner bark of this species was used to produce a tea that was used as an emetic or strong laxative. The tea was also used as an analgesic and to relieve menstrual cramps. The ripe seed coat produces a violet dye. Old *Purshia* stumps produced shredding bark that women would peel off, work with their hands to soften, and use as baby diapers. These were sometimes combined with Juniper (*Juniperus*) bark.

Rose (*Rosa*)

Description: These are shrubs with prickles and leaves that are pinnately divided into 3-11 leaflets. The large, red to pink flowers are borne singly or a few together. The fruits, called hips, are orange, red, or purplish and urn-shaped.

"Quick Key" to the Roses

1. Prickles stout and curved backwards - **R. californica**
1. Prickles slender and straight - **2**

2. Petals 5/8-inch long; leaflets 3/8 - $\frac{3}{4}$-inch long; sepals present on fruit; montane environments - **R. woodsii**
2. Petals 3/8 - $\frac{1}{2}$-inch long; leaflets $\frac{3}{4}$ - $1\frac{1}{2}$ inch long; sepals not present on fruit; of lower elevations - **R. gymnocarpa**

> **California Rose (R. californica)** This shrub is found on dry slopes from 5,000 to 9,000 feet in the San Bernardino Mountains and northward. Flowers from June to August.
> **Wood Rose (R. gymnocarpa)** This species occurs in shaded woods below 6,000 feet from San Diego County northward. Flowers from May to July.
> **Mountain Rose (R. woodsii)** This shrub grows in moist places below 6,000 feet. Flowers May to August.

> **Ecology & Ethnobotany**: The hips are edible raw, stewed, candied, or made into preserves. They are

 high in Vitamin C and also contain Vitamins E, B, and K, beta-carotene, calcium, iron, and phosphorus. There are many other edible parts, besides the fruit. Young *Rosa* shoots in spring make

an excellent potherb, and the roots and stems can be used to make a tea. The petals may be used in salads. The peeled spring shoots can also be nibbled upon. Almost all parts of the plant have been made into a wash or dressing for cuts or sores to coagulate blood. One of the more common methods is to sprinkle fine shavings of de-barked stems into a washed wound. The petals can be used as a dressing. A poultice of leaves can be used to relieve insect stings. In addition, the young leaves can be washed, cut into small pieces, and dried for a hot tea.

Wild Rose Syrup

Gather rose hips in late summer or autumn. Snip off both ends, then cut the hips in half and put them into a pot and barely cover with water. Bring to a boil, cover, and simmer until the hips are tender. Strain off and retain the liquid, then cover the hips with more water and boil for 15 minutes. Strain off and retain the liquid and add it to the first batch. Measure the liquid, then add half as much sugar. Bring to a boil and simmer until the syrup thickens. Pour into sterilized bottles. Serve over pancakes.

Raspberry, Blackberry, Thimbleberry (*Rubus*)

Description: Our species are deciduous shrubs with arching or trailing stems covered with bristles and prickles. The flowers have white petals and the fruit is a coherent cluster of small, 1-seeded drupes (raspberries, blackberries, dewberries, cloudberries, marionberries).

The genus name is derived from the Latin *ruber*, meaning red, in reference to the color of the fruit. This is a large and complicated group taxonomically.

Western Raspberry (R. leucodermis) This straggly shrub with simple leaves has trailing, prickly stems and grows on dry slopes from 4,000 to 7,000 feet in the San Bernardino, San Gabriel, Mt. Pinos, and the Palomar Mountains. Flowers from May to June.

Thimbleberry (R. parviflorus) This shrub with compound leaves grows in open woods below 8,000 feet from San Diego County northward. Flowers from March to August.

Ecology & Ethnobotany: All species produce edible berries. Fossil evidence shows that *Rubus* species have formed part of the human diet from very early times.

Flowers of all species can be added to salads and can be nibbled upon when hiking. The fresh or dried leaves can be steeped for a tea, alone or in herbal blends. Do not use the wilted or molded foliage, as it may be toxic. The young shoots cut just above ground can be peeled and eaten raw or cooked. A tea from the roots was used to

dry runny noses and a tea from the bark was used to stop dysentery. The plants can also provide a uterine astringent, diuretic, laxative, and mild sedative.

Dewberry Juice

"... cover the raw berries with water, crush them, and strain through cheesecloth. Add sweetening to taste. Uncooked juice should be refrigerated and consumed within a day or two."

Burnet (*Sanguisorba*)

Description: The two species in southern California include *S. minor* (Small Burnet) and *S. occidentalis* (Western Burnet), and are annual herbs with pinnately divided leaves. The flowers are clustered in heads and have no petals. The species are found in waste places or moist soils at the lower elevations. The generic name comes from the Latin *sanguis*, meaning blood, and *sorbeo*, meaning to staunch, referring to the herb's ability to stop bleeding.

Ecology & Ethnobotany: The young leaves make a good salad plant, tasting somewhat like cucumbers. The leaves can be chopped and blended or mixed with other herbs as a seasoning. The dried flowers and leaves can be prepared as a tea. The roots are very astringent and a decoction was used in the treatment of internal and external bleeding and dysentery. The brew can also be used as a mouthwash for gum problems.

MADDER FAMILY (Rubiaceae)

The Bedstraw Family consists of approximately 500 genera and 6,000-7,000 species distributed worldwide. About 20 of genera are native to the United States. The family is of economic importance because of coffee, quinine, and many ornamentals.

Bedstraw (*Galium*)

Description: Despite their small flowers, the various species of *Galium* are unmistakable. They are annual or perennial herbs with 4-angled stems and whorled leaves. The small, 4-parted flowers are white or greenish and the fruits are smooth or bristly hairy. They can be found in various habitats from the low to higher elevations.

"Quick Key" to the Bedstraws

1. Whorls of 6 to 6 leaves - *G. aparine*
1. Whorls of 3 to 6 leaves - **2**

2. Leaves linear, prickle-tipped; plant matted - *G. andrewsii*
2. Leaves not linear, prickle-tipped - **3**

3. Plant annual; leaves in a whorl of 2-4; fruit with hooked hairs - *G. bifolium*
3. Plant perennial; leaves in a whorl of 4-6; fruit without hooked hairs - **4**

---

4. Fruit glabrous; wet places - *G. trifidum*
4. Fruit with hairs; dry places - **5**

5. Leaves ¼ to 5/8-inch long; flowers yellowish, drooping; lower leaves not scale-like - *G. hallii*
5. Leaves 1/8 to ¼-inch long; flowers reddish, erect; lower leaves scale-like - *G. parishii*

Phlox-leaved Bedstraw (*G. andrewsii*) This low, matted perennial grows on dry slopes and ridges from 1,000 to 6,500 feet. Flowers from April to June.

Goose Grass (*G. aparine*) This weak stemmed, scrambling, and straggly annual is common on shaded banks below 7,500 feet. Flowers from March to July.

Low Mountain Bedstraw (*G. bifolium*) This is a slender, glabrous annual grows in moist shade in the San Bernardino Mountains from 5,000 to 10,500 feet. Flowers from June to September.

Nodding Bedstraw (*G. hallii*) This shrubby plant is found on dry, rocky slopes and canyons in the San Bernardino mountains to Mount Pinos and westward. It blooms from May to August.

Parish's Bedstraw (*G. parishii*) This is a slender, many stemmed, tufted perennial plant with a woody root is quite common on dry, rocky slopes from 6,000 to 10,000 feet in the San Gabriel Mountains. Flowers from June to August.

Galium (*G. trifidum*) This is a slender perennial with angled stems grows in wet places below 8,000 feet. Flowers from June to September.

Ecology & Ethnobotany: *Galium* is from the Greek gala, meaning milk, referring to the herbs traditional use as a milk coagulant for making cheese. The rennet (a substance that curdles milk in making cheese and junket) for this use was obtained by blending the herb with an equal amount of salt, covering it with water, and then simmering away half of the fluid.

None of the species of *Galium* are known to be poisonous. Although *G. aparine* (Goose Grass) is the most commonly used species, Tilford (1993) believes, as we do, that all other species can be used similarly. The very young leaves and stems can be used as a potherb. The small hairs on the stems make the plant difficult to swallow raw, boiling or steaming, however, does soften them up. If the stems are too fibrous, use only the leaves. Slow roasted until dark brown and ground, the ripe fruit can be used as a coffee substitute.

Medicinally, the plants were used to increase

 urine flow, stimulate appetite, reduce fevers, and remedy Vitamin C deficiencies. It has diuretic, anti-inflammatory, and astringent qualities, and has been used as a lymphatic tonic. A wash made from the plant is said to remove freckles,

whereas a cool tea is reported to cool sunburns. Many species of *Galium* contain asperuloside, which produces coumarin, giving it the sweet smell of new-mown hay as the foliage dries. Asperuloside can be converted to prostaglandins (hormonelike compounds that stimulate the uterus and affect the blood vessels), making *Galium* species of great interest to the pharmaceutical industry.

Dried, the foliage of Bedstraw has been used as a stuffing for mattresses or as a tinder for starting fires. The roots may yield a red dye, but because the roots are thread-like and produce little dye, collecting enough for a strong dyebath would be fairly laborious.

WILLOW FAMILY (Salicaceae)

This family has 2-3 genera and over 500 species distributed worldwide. *Salix* and *Populus* are native to the United States. The family is of little economic importance, except as a source of ornamentals.

Cottonwoods and Quaking Aspen (*Populus*)

Description: These are trees with sticky, resinous leaf buds, and deciduous leaves. Older trees of some species have gray, rough bark; young bark is smooth and whitish. The flowers are borne in catkins that appear before the leaves. They prefer moist soils and, besides *P. tremuloides* are usually along streams. They grow rapidly and are planted for quick shade or wind protection. The soft wood of some species is used for veneers, boxes, matches, excelsior, and paper.

"Quick Key" to the Poplars

1. Leaves round-ovate; $\frac{3}{4}$ - $1\frac{1}{2}$ inch long - *P. tremuloides*
1. Leaves ovate to deltoid; $1\frac{1}{4}$ - $2\frac{3}{4}$ inches long - **2**

2. Leaves ovate; dark green above, paler below - *P. balsamifera*
2. Leaves deltoid; bright green on both surfaces - *P. fremontii*

Black Cottonwood (*P. balsamifera*) This deciduous tree is common along streams below 9,000 feet from San Diego County northward. At the higher elevations the leaves are more lanceolate than ovate or deltoid. Flowers from February to April.

Fremont Cottonwood (*P. fremontii*) This deciduous tree is fairly common in moist areas below 6,500 feet.

Quaking Aspen (*P. tremuloides*) This deciduous tree is recognized by its fluttering leaves. It occurs in moist places along stream banks from 6,000 to 10,000 feet in the San Bernardino Mountains northward. Flowers from April to June.

Ecology & Ethnobotany: The catkins may be eaten raw or boiled in stews and are a source of Vitamin C. The inner bark can also be eaten as a spring tonic, or dried and ground into a flour substitute or extender. The fresh or dried plant can be used in poultices for muscle aches, sprains, or swollen joints. The primary action of *Populus* is that of an analgesic, used topically and internally. It contains varying amounts of populin and salicin, compounds related to early forms of aspirin. The leaves and bark are most effective parts for tea and aid in diarrhea problems. The wood makes for an excellent bow and drill fire set. Cottonwoods are considered to be botanical indicators of water and trappers often used Aspen as bait in beaver sets.

Fire-by-friction

What computers are to the "modern" humans, the ability to start a fire by friction was just as important to "primitive" humans. A quick look at human history reveals fire as the greatest invention of all time.

There are several methods of starting a fire by friction. The drill and hearth method were used by a great many Native Americans. The hearth was a flat piece of wood, usually a softer wood than the drill. It had a small hole reamed in it and a notch leading to the edge of the hearth. The drill was a piece of wood about 2 feet long and a half-inch or less in diameter. Holding the drill vertically and rapidly rotating it between the palms of the hands, embers were formed in the hole. Embers spilled through the notch onto the tinder placed beside the hearth. By cradling the tinder in the hands and gently blowing on the embers, a flame was produced. Depending on craftsmanship, humidity and other factors, a fire could be created in just a few seconds, or never. To help increase the friction, sand grains were sometimes placed in the hole. Also, "points" of ember producing wood were inserted into straight cane-like sticks when appropriate wood could not be found in long straight pieces. The wood these points were made of were more efficient in starting a glowing ember. This method of making fire could be accomplished by a single person, or two people taking turns rotating the drill. Another variation used a "bow" to rotate the drill.

Some woods are better than others in starting a fire. A few choice woods in the southern California mountains include sagebrush (Artemisia tridentata), certain species of willow (Salix), juniper (Juniperus), aspen (Populus tremuloides), and cottonwoods (Populus).

Willow (*Salix*)

Description: Many species of Willow are found in California. They are mostly shrubs with numerous stems. Flowers are in catkins that appear before, with, or after the leaves. Willows generally grow along streams or other moist habitats. Willows root easily and occasionally form dense thickets. They are often planted to reduce stream bank erosion.

"Quick Key" to the Willows

1. Leaves entire - **2**
1. Leaves toothed - **4**

2. Catkin scales yellow; twigs glabrous - *S. lutea*
2. Catkin scales black; twigs pubescent - **3**

3. Capsule hairy; leaves lance-ovate, 5/8 - $1\frac{1}{2}$ inches wide - *S. scouleriana*

3. Capsule glabrous; leaves linear lanceolate, 3/8 - $\frac{3}{4}$-inch wide - *S. lasiolepis*

4. Shrub 6-15 feet tall - *S. lutea*

4. Tree 15 to 45 feet tall - **5**

5. Petioles 1/8 to 3/8-inch long; not glandular - *S. laevigata*

5. Petioles 1/4 to 5/8-inch long; glandular - *S. lucida*

Red Willow *(S. laevigata)* This tree with yellow to red-brown, glabrous twigs grows along streams below 5,000 feet. Catkins appear from March to May.

Arroyo Willow *(S. lasiolepsis)* This tree or shrub with smooth bark and yellowish, pubescent, young twigs are common along stream banks below 7,000 feet. Catkins appear from February to April.

Golden Willow *(S. lucida)* This tree with rough, brown bark and reddish, glabrous twigs grows along stream banks below 8,000 feet.

Yellow Willow *(S. lutea)* This shrub with yellowish-white glabrous twigs occurs in wet places along streams in the San Jacinto and San Bernardino Mountains between 5,000 and 9,000 feet. Flowers from May to June.

Nuttall's Willow *(S. scouleriana)* This small tree or shrub with yellowish, pubescent, young twigs occurs along streams and in moist areas in the San Jacinto Mountains and northward, usually below 10,000 feet. Flowers April to June.

Ecology & Ethnobotany: The young shoots and leaves can be eaten raw. The bitter inner bark can also be eaten raw, although it is better dried and ground into flour substitute or extender. The plant contains salicin which is similar to aspirin and useful as a substitute. Any part of the Willow can be used to produce a tea for use as an aspirin replacement for headache and body pain. The highest concentrations of salicin, however, are found in the inner bark. Because it is not nearly as strong as aspirin, you may have to drink quite a bit of it. The leaves have astringent properties that are effective when placed on wounds and cuts. Bark was chewed as a toothache remedy. Bark, leaves, twigs, and roots produced medicinal teas, powders, washes, and poultices to relieve pain, swelling, infection, bleeding, and many other ailments. Willows, like the Cottonwoods, are botanical indicators of water. The branches of many willow species are very flexible and make them very useful for traps, arrow shafts, and other needs, such as basketry. The bark can also be used as crude cordage.

Willows are an important basketry plant. They are often used as a foundation material and twinning material for twined baskets. Other uses of the wood include framework for dwellings, fish dams and weirs, racks for drying and cooking food, and light hunting bows. Fiber from bark was used for cordage, nets, and clothing. Also, Willows root easily due to the large amounts of indole acetic acid (IAA), a plant hormone in their stems. IAA can be extracted in cold water from one-inch sections of stem and used to induce rooting of other species for transplanting.

SANDALWOOD FAMILY (Santalaceae)

The Sandalwood Family is comprised of partially parasitic herbs. There are about 30 genera and 400 species distributed in the tropics and temperate parts of the world. The most common representative in North America is the genus *Comandra*.

Bastard Toadflax (*Comandra umbellata*)

Description: Bastard Toadflax is a partially parasitic perennial herb with a waxy surface and a rather woody base. The leaves are linear and the flowers are bell-shaped. The fruit is a 1-seeded, berry-like drupe. Bastard Toadflax is common and widespread in shrublands up to the subalpine zone. *Comandra* comes from the Greek *kome*, meaning "tufts of hairs," and *aner*, which means "man," in reference to the stamens. The roots are blue when cut.

Ecology & Ethnobotany: The mature, brown, urn-shaped fruit of Bastard Toadflax may be eaten raw, and is best when slightly green. They were popular with Native Americans because of their sweet taste. The berries, however, are rarely found in sufficient quantities for more than a pleasant tidbit. Consuming too many berries may cause nausea. Strike (1994) indicates that a root

preparation was used to soothe sore, inflamed eyes.

LIZARD'S-TAIL FAMILY (Saururaceae)

The family consists of only five genera and about seven species that occur in southeastern Asia and in the United States and Mexico. In southern California only genus is *Anemopsis* (Yerba Mansa). In general, members are perennial herbs of moist places.

Yerba Mansa (*Anemopsis californica*)

Description: This plant has a flowering stem with one clasping leaf in the middle of the stem and 1 to 3-petioled leaves in the axil. Flowers occur in dense terminal spikes and are subtended by 5 to 8 white to rose streaked bracts. This is a common plant in marshy places below 6,500 feet, occurring into the yellow pine forest. Flowers from march to September.

Ecology & Ethnobotany: Aside from the fact that the seeds of Yerba Mansa were eaten by Native Americans, the species has many medicinal uses. For example, the aromatic peppery roots were peeled, cut, squeezed, boiled and drunk for pleurisy. An infusion of the root was used for stomach ulcers, chest congestion and colds. The tea from the root also provided a general pain reliever. Other uses include an infusion of the bark to wash open sores and drunk for stomach problems. The bark tea was also used as a laxative. Finally, the leaves, after allowing them to be wilted by heat, were applied to swellings. Other medicinal uses include the roots being pounded and used as a poultice for wounds and bruises

SAXIFRAGE FAMILY (Saxifragaceae)

There are about 30 genera and 580 species in this family, found chiefly in cooler and temperate regions of the northern hemisphere. About 20 genera are native to the United States, with more species occurring in the western part of the country. Several species are cultivated as ornamentals. The family name means "breaker of rocks"- many members are found growing in rock crevices. All members of the Saxifrage Family are at least somewhat edible.

Round-leaved Boykinia (*Boykinia rotundifolia*)

Description: This stout perennial has glandular hairy stems. The leaves are basal, round to heart-shaped and shallowly lobed. Flowers occur in an open cluster and are white. Boykinia grows in wet places below 5,500 feet. Flowers in June and July.

Ecology & Ethnobotany: Some Native groups dried the leaves of Boykinia and wore them inside their basket caps and hats for the pleasant fragrance.

Alumroot (*Heuchera*)

Description: In general, *Heuchera* species are perennial herbs with basal leaves. Flowers are small, saucer- to bell-shaped, and greenish, white or pinkish in color. Their various habitats include moist soils and rocky areas up to the alpine zone. The genus is named in honor of Professor J.H. Heucher (1677-1747), custodian of The Botanic Garden in Germany. Many species of Alumroot readily hybridize, making identification of some plants difficult.

Urn-flowered Heuchera (*H. elegans*) This matted perennial grows in the western part of the San Gabriel Mountains between 4,000 to 9,000 feet.

Heuchera (*H. hirsutissima*) This plant is similar to *H. elegans* and grows in the San Jacinto and Santa Rosa Mountains from 7,000 to 11,000 feet.

Parish's Alumroot (*H. parishii*) This species occurs in rocky situations between 4,500 to 11,000 feet in the San Bernardino mountains.

Ecology & Ethnobotany: The leaves of all species in southern California are edible, although they are not choice. They have a sour taste because of the high tannin content. Therefore, the leaves should be boiled or steamed. Since they are rather tough, we found them to be more palatable if chopped and added to soups or salads.

Heuchera is said to be one of the strongest astringents due to their high tannin content. Tilford (1997) indicates that the root of these plants may contain as much as 20 percent of its weight in tannins. Tannins tend to shrink swollen, moist tissues. Therefore, they are also gastrointestinal irritants and have been known to cause kidney and liver failure. Ingestion of the plant should be in moderation. Otherwise, the pounded, dried roots of many species have been used as a poultice that stops bleeding and promotes healing when applied to cuts and sores. The raw root, eaten in small amounts, has been used as a cure for diarrhea. A tea from the roots can also be used as a gargle for sore throats. The powdered roots have been used as an antiseptic.

Alumroots are also commonly used as mordants, substances that make natural dyes colorfast. The Alumroot of choice, however, occurs in western deserts near sulphur springs, but satisfactory substitutions probably do exist in California.

Dyes, Dyeing, and Mordants

Plants have provided humans with color for uncounted ages. In the last 5,000 to 7,500 years, humans have learned to transfer some of Nature's colors to cloth, paper, wood, leather, and wool. In comparison, natural dyes do not last as long as the synthetics of today, but when your living primitively, natural dyes are readily available. Most vegetable dyes are prone to fading and need some added treatment to become color-fast. This process is called mordanting, that is, treating the material to be dyed with other substances that serve to fix the color. Basically, what mordanting and dyeing involves soaking (or boiling or simmering) the material in water with the dissolved mordant. This is dependent on dye source and material to be dyed. After a prescribed time, the material is rinsed and allowed to dry. Dye bath is prepared by soaking the chopped or crushed plant material in water overnight and boiling until the appropriate color is extracted. Plant material is strained out and water is added to make 4-5 gallons of lukewarm dye bath, to which 1 pound of fabric is added. After dyeing and stirring as long as necessary for color desired, the dyed material is passed through a series of rinses, each a little cooler than the previous one, until the rinse water remains clear. After drying, the dyed material is ready for use.

Woodland Star (*Lithophragma*)

Description: These perennial herbs have slender bulbet-bearing rootstalks. There are approximately 10 species in western North America. The genus name is from the Greek, *lithos* (rock) and *phragma* (fence). Incidentally, generic names that end in "*phragma*" are considered of neuter, not feminine, gender.

Woodland Star (*L. heterophyllum*) This perennial grows on shaded slopes below 6,500 feet. Flowers from March to June.

Lithophragma (*L. tenellum*) This perennial grows in dry areas from 5,000 to 7,500 feet in the San Gabriel and San Bernardino mountains northward. Flowers from May to June.

Ecology & Ethnobotany: We found no recorded uses for southern California mountain species, but the root of a species in the Sierra Nevada, *L. affinis*, was chewed by the Maidu, Mendocino, and Yuki Indians of California to treat stomach ailments and colds.

Brook Saxifrage (*Saxifraga odontoloma*)

Description: This perennial with basal leaves and a leafless flowering stem has round to kidney-shaped leaves. The white flowers occur in a few-flowered open and spreading inflorescence. This species grows on moist stream banks from 6,500 to 11,200 feet in the San Bernardino Mountains. Flowers from July to August. The generic name is from Latin *saxum* meaning rock and *frangere*, meaning to break. It alludes to the species rocky habitat. Herbalists once used some species in a treatment for "stones" in the urinary tract.

Ecology & Ethnobotany: The genus as a whole is regarded as a safe group of plants. The leaves can be used fresh or in stews and are high in Vitamins A and C. Richard Scott (Central Wyoming College) shared with us this favorite back country salad of his:

"...*when backpacking, we always carry a little bottle of vinegar, one of oil, and one of dried or powdered salad dressing. In fact, an excellent salad is based on saxifrage, particularly S. arguta (Brook Saxifrage) leaves with a few alpine sorrel leaves (Oxyria digyna), a couple of Indian Paintbrush inflorescences (Castilleja miniata or C. rhexifolia), and a couple Columbine flowers (Aquillea caerulea)."*

In China some species were used in the treatment of nausea and ear infections. In our area, there is little documentation regarding medicinal uses.

SNAPDRAGON FAMILY (Scrophulariaceae)

The Snapdragon Family has about 220 genera and 3,000 species distributed worldwide. About 40 of the genera are native to the United States. The family is of economic importance because of the cardiac glycosides derived from *Digitalis* (Foxglove) and the many fine ornamentals.

White Snapdragon (*Antirrhinum coulterianum*)

Description: This erect annual grows up to 4 feet tall, and has glabrous stems and twisted, upper branchlets. The leaves are alternate, lanceolate in shape. The lowermost leaves are slightly shorter and opposite. Flowers occur in a long, dense raceme, and are subtended by narrow bracts. The corolla is white and resembles a snapdragon. White Snapdragon is common below 5,000 feet in dry areas after burn. Flowers April to July.

Ecology & Ethnobotany: A related species, *A. nuttallianum*, was used as a decoction to relieve colds.

Paintbrush (*Castilleja*)

Description: This is a large genus found primarily in western North America that contains many species found southern California. The genus is easily recognized, but many species are notoriously difficult to identify. They are perennials with deeply lobed to entire leaves. The flowers are subtended by colorful leaf-like bracts. Some Paintbrushes are partial root parasites found in various habitats up to the alpine zone. The genus name honors the Spanish botanist Domingo Castillejo.

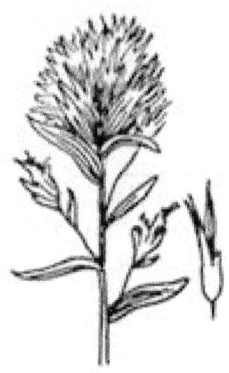

"Quick Key" to the Paintbrushes

1. Herbage gray-green or white-woolly - **2**
1. Herbage green; glabrous or short pubescent - **3**

2. Herbage white-woolly; leaves linear with smaller leaves in axils; corolla less than 1 inch long - *C. foliolosa*
2. Herbage gray-green; leaves lanceolate; corolla 1 - 1½ inches long; plant of desert mountains - *C. angustifolia*

3. Leaves with fascicles of smaller leaves in the axils - *C. affinis*
3. Leaves without fascicles of smaller leaves in the axils - **4**

4. Herbage glabrous below the flowering cluster - *C. miniata*
4. Herbage glandular pubescent - **5**

5. Calyx lobes 1/8-inch long; Ventura County and north - *C. applegatei*
5. Calyx lobes 1/16-inch long; Los Angeles County and south - *C. angustifolia*

Paintbrush (*C. affinis*) This erect perennial grows on dry, wooded, or brushy slopes. Flowers from March to May.

Desert Paintbrush (*C. angustifolia*) This is a grayish green perennial is common on dry, brushy slopes between 2,000 and 10,000 feet on the desert side of the mountains. Flowers from April to August.

Paintbrush (*C. applegatei*) This glandular, pubescent perennial is common in dry rocky areas between 2,000 to 11,000 feet on Mt. Pinos and northward. Flowers from May to August.

Woolly Paintbrush (*C. foliolosa*) This white woolly perennial grows on dry, rocky slopes below 5,000 feet. Flowers from March to June.

Great Red Paintbrush (*C. miniata*) This erect, glabrous, or slightly pubescent perennial is common in wet places and flowers from May to September.

Ecology & Ethnobotany: Many, if not all of the species have flowers and bracts that can be eaten raw. The seeds of some species were gathered, winnowed, dried, and stored for winter use. In winter they were parched, pounded and eaten dry. The plants, however, absorb selenium from the soil and so should be taken in moderation. Symptoms in humans of selenium poisoning will vary with the amount and form ingested, but may include difficulty in breathing, excessive urine production, loss of appetite, mental depression, a weak and rapid pulse, blurry vision, digestive upset, and eventually coma and death.

Collinsia (*Collinsia*)

Description: These are annual plants that are often glandular and brown staining. The leaves are opposite and the flowers are 5-lobed appearing pea-like. The genus name honors Zaccheus Collins, a Philadelphia botanist that lived from 1764 to 1831.

Collinsia (*C. callosa*) This stout annual grows in dry places between 3,000 and 7,500 feet in the San Bernardino, San Gabriel, and Tehachapi Mountains. Flowers from April to June.

Blue Lips (*C. parviflora*) This annual plant is slightly pubescent and grows in moist, shaded areas from 2,500 to 11,150 feet throughout the area. Flowers from April to July.

Blue-eyed Mary (*C. torreyi*) This erect annual grows between 7,000 and 11,000 feet in the San Gabriel and San Bernardino Mountains. Flowers from June to August.

Ecology & Ethnobotany: The leaves of a related species, *C. heterophylla*, was used by the Maidu as a poultice for bites from insects or snakes.

Bird's-beak (*Cordylanthus nevinii*)

Description: This is a slender annual, growing up to 20 inches tall that has bristly hairs on the stem and leaves. The upper leaves are linear. Flowers number 1 to 3 in a cluster, with each cluster subtended by green, 3 lobed bracts. The calyx is slightly purplish, and 2 toothed. The corolla is purplish and 2 lipped. Bird's Beak grows on dry slopes between 5,000 and 8,000 feet in the San Gabriel, San Bernardino, and San Diego Mountains. Flowers from July to September.

Ecology & Ethnobotany: Strike (1994) indicates that *Cordylanthus* was used by the Luiseno Indians in California as an emetic.

Monkeyflower (*Mimulus*)

Description: Many species of Monkeyflower can be found in southern California. They are annual or perennial herbs with opposite leaves. The flowers flare at the mouth to form 5 lobes, 2 which form the upper lip and 3 lobes that form the lower. The perennial species occur in permanently wet soil, while the annual are found in vernally moist habitats.

Monkeyflower (*M. biglovii*) This glandular, pubescent annual is common on the desert side of the mountains below 7.000 feet. Flowers from March to June.

Scarlet Monkeyflower (*M. cardinalis*) This hairy and slightly glandular perennial is quite common in wet places such as along streams and creeks, below 8,000 feet. Flowers from April to October.

Monkeyflower (*M. fremontii*) This glandular, hairy annual grows in dry, disturbed areas below 7,000. Flowers from April to June.

Seep Monkeyflower (M. guttatus) This perennial is common along streams and in wet places. Can be found throughout the area below 10,000 feet. Flowers from March to August.

Monkeyflower (M. johnstonii) This is a glandular, pubescent annual grows on dry, gravelly areas between 4,000 and 7,000 feet from Santa Barbara County south to San Diego County. Flowers from May to July.

Bush Monkey Flower (M. longiflorus=M. aurantiacus) This low shrub is found growing in rocky places up to 7,500 feet in the mountains. Flowers April to July.

Musk Flower (M. moschatus) This weak stemmed perennial grows in wet places below 7,500 feet. Flowers from June to August.

Monkeyflower (M. nasutus = M. guttatus) This glabrous to slightly pubescent annual grows in wet, sandy, or gravelly places below 7,500 feet. Flowers from March to August.

Downy Monkeyflower (M. pilosus) This is a low annual is common in moist, sandy or gravelly places, below 8,500 feet. Flowers from April to September.

Primrose Monkeyflower (M. primuloides) This low perennial grows from Riverside County northward at elevations below 11,000 feet. Flowers from June to August.

Monkeyflower (*M. tilingii*) This is a glabrous to pubescent perennial grows on wet banks between 6,400 and 11,000 feet in the San Jacinto and San Bernardino Mountains. Flowers from July to September.

Ecology & Ethnobotany: The young stems and leaves of *M. guttatus* (Seep Monkeyflower) have been used as salad greens. Sometimes, leaves were burned and the ash used a salt. Weedon (1996) indicates that the young herbage of *Mimulus* species may be eaten in salads, and that they grow bitter with age, but remain edible. For example, the leaves of *M. primuloides* (Primrose Monkeyflower) were eaten by the Maidu Indians in California and the young plants of *M. moschatus* (Musk Monkeyflower) were boiled and eaten by some California Natives.

There are also reports of some Native Americans making a tea from monkey flowers stems, leaves, and flowers as a treatment for kidney or urinary problems and to cure diarrhea. Monkeyflower leaves and stems were used externally to poultice wounds or internally to reduce fevers. *Mimulus* roots were used to treat fevers, dysentery, and diarrhea and to curtail hemorrhages. The raw leaves and stems can be applied to burns and wounds as a poultice.

Owl's Clover (*Orthocarpus*)

Description: Members of this genus are semi-parasitic slender plants with sessile flowers that are subtended by a large, colorful bract. The tubular flowers are 2-lipped with 2 upper lobes united to form a small hood-shaped galea. The 3 lower lobes are partly united and nearly as long as the galea. Owls Clovers resemble Paintbrushes (*Castilleja*), but can be distinguised by the annual growth form and by having a lower corolla that nearly equals the galea. San Bernardino Mountains Owl's Clover (*O. lasiorhynchus* = *Castilleja lasiorhynca*) is a slender stemmed annual common in meadows between 4,000 and 7,500 feet from the San Bernardino Mountains south to Cuyamaca. Flowers in June and July.

      Ecology & Ethnobotany: The foliage of a related species *O. luteus* was used by some Natives to dye small skins and feathers a reddish tan color.

Lousewort (*Pedicularis*)

      Description: These are perennial, partially parasitic herbs with toothed or pinnately divided leaves. The flowers are subtended by leaf-like bracts and occur in a spike-like inflorescence. The tubular flower is strongly 2 lipped at the mouth. They are found in open, dry or moist habitats, including meadows up to the alpine zone. The genus name comes from the Latin *pediculus* (little louse) alluding to the superstition that livestock that ate the plants would suffer an infestation of lice.

      Indian Warrior (*P. densiflora*) This perennial herb grows in dry slopes up to 6,000 feet elevation.
      Pine Lousewort (*P. semibarbata*) This perennial herb is found in the pine forests between 5,000 to 11,000 feet.

      Ecology & Ethnobotany: Tilford (1997) indicates that as long as Lousewort is not attached to an unpalatable host, the fleshy roots can be prepared and eaten in moderation. He does not, however, describe methods of preparation. He also states that the leaves and stems of some species may be steamed or boiled as potherbs, but this is not recommended.

Moore (1979) describes *Pedicularis* as an effective sedative for children and tranquilizer for adults. The whole stalk, when dried and prepared as a tea, acts as a mild relaxant, quieting anxiety and tension. The fresh or dried plant is a vulnerary for minor injuries, with mild astringent and antiseptic properties.

Warning: Because of the sedative nature of these plants, the potential toxic alkaloids, and the host species it may be attached to, ingestion of this plant is not recommended.

Penstemon (*Penstemon*)

Description: There are many species of Penstemon. In general, they are perennial herbs with opposite leaves. The flower is strongly to indistinctly 2-lipped at the mouth with a 2 lobed upper lip and a lower lip with three lobes. There are four anther-bearing (fertile) stamens and a single sterile stamen that is often hairy at the tip. The fruit is a many seeded capsule. Penstemons occur in dry or moist meadows or forest openings up into the alpine zone. The genus name is from the Greek *pete*, meaning five, and *stemon*, meaning thread, referring to the slender fifth stamen.

Scarlet Bugler (*P. centranthifolius*) This gray green, glabrous perennial is common in dry, sandy areas below 6,500 feet. Flowers from April to July.

Eaton's Firecracker (*P. eatonii*) This glabrous perennial is found on dry, gravelly slopes below 8,000 feet, mostly on the desert side of the mountains. Blooms from March to July.

Beard Tongue (*P. grinnellii*) This branched perennial plant grows on dry, gravelly slopes between 4,500 and 9,500 feet in the San Gabriel and the Santa Rosa Mountains. Flowers from May to August.

Penstemon (*P. labrosus*) This bright green perennial grows on dry slopes from 5,000 to 10,000 feet from Ventura County south to San Diego County. Flowers from July to August.

Scarlet Penstemon (*P. rostriflorus*) This is a shrub that occurs on dry slopes from 5,000 to 10,700 feet. Blooms June to August.

Ecology & Ethnobotany: As a topical astringent, pureed or juiced, the plants can be used as a general dressing for minor irritations of the skin (e.g., insect bites). Penstemon oil is a good addition to an all-purpose salve.

California Figwort (*Scrophularia californica*)

Description: This coarse, tall perennial grows up to 6 feet tall, and has 4 angled stems. The leaves are opposite and ovate in shape. Flowers are small and occur in an open inflorescence. The corolla is 2 lipped and maroon. Figwort is common in moist places in the woods. Flowers from February to July. There are four functional stamens and the fifth one is reduced to a small knob on the corolla.

Ecology & Ethnobotany: A strong tea made from the plant can be applied to fungal infections of the skin (e.g., athletes' foot), and can help eczema, rashes, burns, and hemorrhoids.

Mullein (*Verbascum thapsus*)

Description: The white or yellow, saucer-shaped flowers occur in a spike-like inflorescence. The species can be found in disturbed places up to the subalpine zone.

Ecology & Ethnobotany: The leaves of Mullein are said to be edible when eaten in small quantities and cooked. Because of their woolly texture, however, we have found the plants to be undesirable.

Native Americans smoked the dried leaves of Mullein for asthma and sore throat. A tea from leaves was used to treat colds and the flowers contain an oil that has been used for earaches. The first-year leaves make a soothing decoction for coughs and sore throats. Boil leaves in water for ten minutes, then strain the liquid through a cheesecloth to remove the tiny hairs. The leaves can also be used as poultices applied locally to hemorrhoids, sunburn, and inflammations. The dried stalks are ideal for use as hand-drills to start fires. The flowers and leaves

produce yellow dye; as a toilet paper substitute, the large fresh leaves are choice.

Caution: Mullein does contain coumarin and rotenone, two substances that may be toxic in large quantities if ingested. Also, the seeds are not recommended for consumption.

American Speedwell (*Veronica americana*)

Description: This glabrous and slightly succulent perennial has a stem that creeps at the base,

 and is 4 to 40 inches long. Leaves are opposite, lanceolate to ovate in shape, and toothed. Flowers are in axillary racemes. The corolla is 4 lobed, slightly irregular, and violet-blue to lilac. There are 2 stamens. Brooklime grows in wet places along streams from sea level to 10,500 feet. Flowers from May to August.

The mature fruit is often necessary for identification. The genus may be named after Saint Veronica, who was said to have wiped the face of Christ on route to his crucifixion. The name may also have arisen to its beginning as a medicinal herb, after a shepherd observed and injured deer heal its wounds by rolling in and eating the herb. The shepherd then reported this to his sick king. The king became well after trying the and showered the shepherd with riches.

Ecology & Ethnobotany: The leaves and stems of all species, when collected during the spring and early summer, can be eaten like watercress, added to salads, or prepared as potherbs. The taste of the various species ranges from spicy to bitter to bland depending on personal taste. The plants also contain moderate amounts of Vitamin C and were once used to prevent scurvy. The leaves and stems can also be steeped as a tea. Care should be taken to avoid plants growing in polluted waters. If desired, the addition of halazone tablets or chlorine bleach to the wash water may kill harmful microbes. After flowering, it is better to boil the plants to eliminate the bitterness.

Medicinally, the plants were used mainly as an expectorant for respiratory problems. Infusions were also used in hair conditioning rinses and skin cleaning herbal steams, and as an ingredient in massage oils and ointments that were added to baths for a soothing soak. Pojar and MacKinnon (1994) indicate that *V. americana* has been used for centuries to treat urinary and kidney complaints and as a blood purifier. The leaf juice from *V. serpyllifolia* (Thymeleaf Speedwell) has been used for earaches, its leaves were poulticed for boils, and a tea was used for chills and coughs.

NIGHTSHADE FAMILY (Solanaceae)

There are about 85 genera and 2,300 species in the Nightshade Family. About 13 genera are indigenous to the United States. The family is noted for its edible, poisonous, and medicinal plants. Some of the economically important plants include tomato, potato, chili, tobacco, and eggplant. Medicinally, two alkaloids, atropine (belladonna) and scopolamine which can be deadly in large amounts are obtained from this family. Eye doctors today use atropine in very exact small, doses as an effective dilator when examining eyes.

Jimson Weed (*Datura wrightii*)

Description: This is a coarse, erect annual or perennial plant, growing up to 40 inches tall, and having a strong odor and narcotic qualities. Leaves are alternate, ovate, unequal at the base, and with wavy margins and petioles. Flowers are large, solitary, and in the forks of branching stems. The calyx is tubular-shaped and 5 toothed. The corolla is funnel-shaped, white, and purple to violet. Flowers are pleated in the bud and open in the evening. Fruit is a prickly, globose capsule with spines. Jimson Weed grows on sandy or gravelly open areas, generally below 6,000 feet. Flowers from April to October.

Ecology & Ethnobotany: Extracts from these plants are narcotic and, when improperly prepared, are very lethal. *Datura* is also known as "hell's bells" among drug users because of the popular misconception that the

plants provide worthwhile hallucinogenic experiences. Unfortunately, who experimented with *Datura* as a drug died as a result. Those that did survive realized that it was not worth the try. Narcotic properties have been known since before recorded human history and they once figured importantly in the religious ceremonies of many Native Americans.

Although potentially deadly, the plants do have valuable medicinal values. There are several alkaloids including atropine and hyoscyamine that have been used in scientifically refined antispasmodic drugs. *Datura* also contains scopolamine, an antivertigo compound commonly used to treat motion sickness and other conditions involving disequilibrium. The concentrations of these compounds vary widely among plant parts, species, and localities, making *Datura* highly unreliable for internal use, even in trained hands.

Warning: These plants should never be taken internally for any reason. They contain dangerous alkaloid compounds that could easily kill a person.

Boxthorn (*Lycium andersonii*)

Description: This thorny shrub may grow to 9 feet tall. The leaves are clustered, drought-deciduous, and fleshy. The flowers are lavender in color and arise from between the forks of branches. This species grows in woodlands and scrub habitat up to 6,000 feet.

Ecology & Ethnobotany: The plants produce large crops of berries that were an important food item for California Natives. They were eaten raw or dried in the sun like raisins. The dried berries can be boiled into mush or ground into flour and mixed with water. The roots can also be dried, pounded, baked and eaten.

Coyote Tobacco (*Nicotiana attenuata*)

Description: This annual plant grows up to 40 inches tall, and has glabrous to glandular pubescent stems. The leaves are large, ovate to ovate lanceolate in shape, and the upper leaves are narrowed. Flowers occur in a terminal raceme, and the calyx is 5-cleft. Corolla is white, tubular or funnel shaped. The flowers close in the sun. Coyote Tobacco grows in disturbed places below 8,500 feet. Flowers from May to October.

Ecology & Ethnobotany: All species of Nicotiana contain the highly toxic alkaloid nicotine. An effective insecticide against aphids can be prepared by steeping Tobacco leaves in water and spraying the solution on affected parts of the plant.

Purple Nightshade (*Solanum xantii*)

Description: This perennial herb has stems that grow up to 16 inches tall. The stems are densely, gray hairy; and may form mats on the ground. The leaves are ovate in shape. The corolla is 5 angled, deep violet, and the anthers are fused around the style. Fruit is a greenish berry. Purple Nightshade grows on dry places from 5,000 to 9,000 feet in the San Bernardino Mountains and north. Flowers from May to September.

This is a highly diverse genus comprising more than a thousand species worldwide. *Solanum* probably comes from the Latin "*solamen*," which means quieting, referring to the sedative properties of some species.

Ecology & Ethnobotany: The most poisonous part of the plant is the unripe fruit, but the stems, leaves, and roots are also dangerous. In fact, the alkaloid content in the plants decreases in this order: unripe fruit-leaves-stems-ripe fruit. Solanine is the predominant glyco-alkaloid, but others may be present. Solanine is highly toxic and can cause death, but the degree of toxicity varies among and within species. It is reported that cooking destroys the solanine, making the ripe fruit edible.

The berries of *S. nigrum* (Black Nightshade) were formerly used as a diuretic (Foster and Duke 1990). Native Americans of the Southwest used the crushed berries to curdle milk for making cheese and they have also been used in various preparations for sore throat and toothaches.

Solanum dulcamara (Climbing Nightshade), when used correctly and in appropriate dosages, is said to be useful in the treatment of skin disorders, rheumatism, and bronchitis. Recent studies have shown that this plant possesses anti-cancer qualities.

Caution: There are references listing the berries of some species of *Solanum* as edible. However, it is our recommendation that none of the plants that occur in southern California be consumed in any manner.

CACAO FAMILY (Sterculiaceae)

This family is comprised of herbs, shrubs, and sometimes woody vines. The leaves and stems for the most part have stellate hairs. The family is distributed throughout the Tropics and Subtropics. There are at least five genera in the United States. The most famous member of this family is *Theobroma cacoa* (Cocoa Tree), which is native to tropical America. This plant has large reddish yellow fruits, and the seeds are the source of chocolate and cocoa. Another species, *Cola acuminata* (Cola Nut), found in Africa is the source for cola drinks.

Flannel Bush (*Fremontia californicum*)

Description: This is an evergreen shrub, 5 to 20 feet tall. The leaves are simple, alternate, round-ovate, palmately 3-lobed, dark green and slightly pubescent above and yellowish, densely pubescent below. The showy flowers are bright yellow. There are no petals, and the calyx is petal-like with 5 lobes. The 5 stamens are attached to the style of the pistil. The fruit is a bristly hairy, ovoid capsule. Flannel Bush grows on dry slopes between 3,000 and 6,000 feet from San Diego County north. Blooms from May to June.

Ecology & Ethnobotany: Medicinally, an infusion of the inner bark was taken as a physic. The inner bark was also used as a poultice for sores, and when soaked in water, the inner bark works as a purgative.

The fiber from the outer bark can be twisted into cordage, rope, or nets. The stems were used for cradleboard frames, and the smaller branches were used to make bows and arrows. One Native American tribe, the Yokuts (San Joaquin Valley area), considered *Fremontia* a "one-stop" weapons store. The bark was stripped and twisted into good bowstring, whereas the large branches were split into staves for bows and small straight branches were used for arrows.

STYRAX FAMILY (Styraceae)

This family occurs in the warmer parts of the Americas, in the Mediterranean, and in eastern Asia.

Snowdrop Bush (*Styrax officinalis*)

Description: This is a deciduous shrub with grayish twigs grows 3 to 12 or more feet tall. The entire leaves are simple, alternate, round ovate, and is pubescent above and below. The white, showy flowers are fragrant, and occur in drooping terminal clusters. The corolla is variable in the number of lobes, but is usually 6. The flowers resemble orange blossoms. The fruit is globular and dry. Snowdrop Bush grows on mountain slopes and in canyons at lower elevations below 5,000 feet in San

Bernardino and San Diego counties. Flowers from April to May.

Ecology & Ethnobotany: The Maidu used sap from this plant as an antiseptic and as an expectorant.

ELM FAMILY (Ulmaceae)

Members of this family are trees or shrubs with simple, alternate leaves. The flowers lack petals and the fruit is a samara, nut, or drupe. There are approximately 15 genera and 200 species in this family distributed through the temperate and subtropical regions of the Northern Hemisphere. *Ulmus, Planera,* and *Celtis* are native to the United States. The family is of economic importance as a source of wood.

Net-leaved Hackberry (*Celtis reticulata*)

Description: This is a small trees or shrubs with leaves that are ovate to lanceolate in shape, with entire to serrate edges. The fruit is a drupe, and the plants can usually be found growing along streams or on dry canyon slopes at the lower elevations between 2,800 to 5,500 feet in the San Bernardino mountains and Peninsular Ranges.

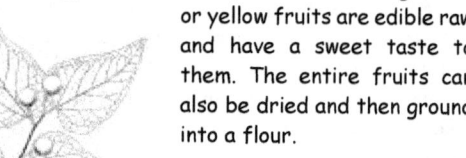

Ecology & Ethnobotany: The small orange, red, or yellow fruits are edible raw and have a sweet taste to them. The entire fruits can also be dried and then ground into a flour.

NETTLE FAMILY (Urticaceae)

There are about 45 genera and 550 species found in the Nettle family. Six of the genera are native to the United States and are of little economic importance.

Stinging Nettle (*Urtica dioica*)

Description: This is an annual or perennial herb with stinging hairs. The flowers are numerous, small, and clustered on drooping branches at the base of the leaves. Stinging Nettles can be found along roadsides, streams, in moist areas and waste places in the low to middle elevations. Stinging Nettle is an indicator of good soil conditions. Many people consider Stinging Nettles, for obvious reasons, as obnoxious weeds.

Ecology & Ethnobotany: One of the first things a person learns about Stinging Nettles is their stinging effect. The intense burning and itching or stinging of the skin may persist for a length of time. If you look closely at the hairs, you will see a hypodermic mechanism consisting of a very fine capillary tube with a bladder-like base that is filled with chemical irritant. When brushed against, a minute spherical tip breaks off uncovering a very sharp-pointed tip that easily penetrates the skin. The chemical is forced into the skin through the tube as the hair bends and constricts the bladder-like base. Therefore, Stinging Nettles should be collected with gloves.

The young stems and leaves of Stinging Nettle are edible after boiling, and are very delicious as a spinach substitute. Boiling the leaves destroys the formic acid found in the hairs. The leaves are high in vitamins A, C, and D, the latter of which is rare in plants. The roots are also edible after they have been roasted. A tea made from the leaves is said to have astringent and diuretic qualities, and has been used for internal bleeding and nosebleeds. Native Americans learned of Stinging Nettle as a food from early European travelers and settlers, and possibly from Chinese immigrants.

The older stems become fibrous, which reduces their edible qualities, but allows them to be used to produce strong cordage. The older leaves also contain cystoliths that can irritate the kidneys. A yellow dye may be obtained by boiling the roots.

A tea brewed from *Urtica* was said to relieve chest colds and internal pains. As a poultice, *Urtica* was used for headaches.

VERBENA FAMILY (Verbenaceae)

Approximately 75 genera and 3,000 species occur worldwide, of which 14 genera are native to the United States. The family is of economic importance because of the highly prized teak wood (*Tectona grandis*) and a number of other ornamentals.

Verbena (*Verbena*)

Description: Two species can be found in the southern California mountains. Members of this genus are perennials with opposite, toothed leaves. The tubular flowers with flaring lobes each have a subtending bract. The fruit is a cluster of 4 nutlets.

Desert Vervain (*V. goodingii*) This glandular, hairy perennial grows on dry slopes in the desert mountains from 4,000 to 6,500 feet. Flowers from April to June.

Western Verbena (*V. lasiostachys*) This is a much-branched perennial with long, hairy stems grows on dry or moist slopes below 8,000 feet. Flowers from May to September.

Ecology & Ethnobotany: The seeds of a related species, *V. hastata* (Blue vervain), may be gathered, roasted, and ground into a bitter tasting flour. Leaching the flour may remove the bitter taste. A tea from boiled leaves can be used for a stomachache, and a tea from roots was used to clear cloudy urine. Moore (1979) says that the plant is used as a sedative, diaphoretic, bitter tonic, and mild coagulant. It promotes sweating, relaxes and soothes, settles the stomach, and gives an overall feeling of relaxed well-being.

VIOLET FAMILY (Violaceae)

The Violet Family has approximately 16 genera and 850 species distributed worldwide. Two genera are native to the United States. The family is of little economic importance other than a source of ornamentals as many species of *Viola* are cultivated. *Viola* is the classical name for violets.

Violet (*Viola*)

Description: There are many species of Violets in California. In general, they are low-growing, perennial or annual herbs. The leaves are spade-shaped and basal. The flowers occur singly on the ends of stems and have five petals. There are 2 upper and 2 lateral petals, and 1 lower petal that is prolonged into a nectar holding pouch at the base of the flower. Most species also have small, self-fertilizing flowers that do not open. Violets can be found in meadows and open forests from the foothills to above timberline.

"Quick Key" to the Violets

1. Flowers violet or white - **2**
1. Flowers yellow with dark veins and dark on back of upper petals -**3**

2. Flowers violet - *V. adunca*
2. Flowers white - *V. macloskeyi*

3. Leaves simple, not lobed or divided - *V. purpurea*
3. Leaves compound or palmately divided - **4**

4. Leaves 2 times pinnately divided into linear lobes - *V. douglasii*
4. Leaves palmately divided - **5**

5. Stems only a little above ground; leaves basal - **V. sheltonii**
5. Stems 4 to 12 inches long; leaves at the top of stem - **V. lobata**

Western Dog Violet (V. adunca) This slightly pubescent perennial grows on moist banks and at the edge of meadows from 5,000 to 8,000 feet. Flowers from March to July.

Douglas Violet (V. douglasii) This is a pubescent perennial grows in open grassy areas in the mountains from 3,500 to 7,500 feet. Flowers from March to May.

Pine Violet (V. lobata) This is an erect, slightly pubescent perennial grows from 1,000 to 6,500 feet. Flowers from April to July.

Macloskey's Violet (V. macloskeyi) The white flowered violet occurs in wet meadows, seeps, and along streambanks from 3,000 to 10,500 feet in the San Bernardino and San Jacinto mountains and northward.

Mountain Violet (V. purpurea) This pubescent perennial grows on dry slopes below 6,000 feet. Flowers from April to June.

Violet (V. sheltonii) This glabrous perennial grows in the shade of open woods or in brushy areas from 2,500 to 8,000 feet. Flowers from April to July.

Ecology & Ethnobotany: The leaves, buds, and flowers of possibly all species are edible raw or cooked, with some being more palatable than others; the leaves make a good tea. Adding the leaves to soups make them

thicker. Violets are high in Vitamin C and beta-carotene. Collect the plants by leaving the roots intact. Since many species can reproduce vegetatively, you will probably not inhibit next year's growth significantly. Many naturalists indicate that all violets are safe for consumption, but there are some experts that insist some yellow species may be somewhat purgative. All species have the tendency to be slightly laxative, so proceed slowly. The flowers have also been candied or made into jellies and jams.

Violet salve can be made by simmering the entire herb in lard. It was a famous remedy for skin inflammations and abrasions. Violets are also emollients (soften the skin) and are an excellent ingredient in lotions such as night creams. Flowers and leaves of some species have been used in various herbal remedies as poultices and laxatives and to relieve cough and lung congestion.

MISTLETOE FAMILY (Viscaceae)

These are parasitic shrubs found on the branches of a variety of trees such as oaks, alders, conifers, and cottonwoods. Leaves are opposite and somewhat leathery. The flowers are small, comprised of 2-4 tepals (petals and sepals not differentiated) inserted on a cup-like receptacle. The fruit is a drupe or berry.

There are approximately 11 genera and 450 species in this family distributed worldwide. In the United States, *Arceuthobium* and *Phoradendron* are native.

Dwarf Mistletoe (*Arceuthobium*)

Description: These species grow as parasites on coniferous trees and have opposite leaves. The flowers are found in the axils of the leaves. The pulp of the berry is mucilaginous which glues the seed to whatever it touches (to eventually germinate on the trees). These Mistletoes are found on a variety of conifers. The genus name is from the Greek *arkeuthos*, meaning juniper, and bios, meaning life, as the plant is often a parasite on junipers.

Fir Dwarf Mistletoe (*A. abietinum*) On White Fir (*Abies concolor*).

Lodgepole Pine Dwarf Mistletoe (*A. americanum*) On Lodgepole Pine (*Pinus contorta*)

Sugar Pine Dwarf Mistletoe (*A. californicum*) On Sugar Pine (*Pinus lambertiana*)

Western Dwarf Mistletoe (*A. campylopodum*) On Jeffrey Pine (*Pinus jeffreyi*) and Ponderosa Pine (*P. ponderosa*).

Limber Pine Dwarf Mistletoe (*A. cyanocarpum*) On Limber Pine (*Pinus flexilis*).

Pinyon Dwarf Mistletoe (*A. divaricatum*) On Pinyon Pines (*Pinus edulis* and *P. monophylla*)

Foothill Pine Dwarf Mistletoe (*A. occidentale*) On Coulter Pine (*P. coulteri*)

Ecology & Ethnobotany: Strike (1994) indicates that *A. campylopodium* mixed with the pitch of *Pinus edulis*, was used to cure coughs and colds and to relieve the associated aches. Additionally, the smoke from burning mistletoe relieved colds and coughs and a decoction was used as a contraceptive.

A decoction of Lodgepole Pine Dwarf Mistletoe was taken for lung and mouth hemorrhages, as well as for tuberculosis. A decoction of Foothill Pine Dwarf Mistletoe was taken for stomachaches.

Juniper Mistletoe (*Phoradendron juniperinum*)

Description, Ecology & Ethnobotany: This species is usually found growing on Junipers (*Juniperus*). The berries were eaten fresh by Cahuilla. Medicinally, a poultice from the mashed berries were used on wounds to aid in healing, and a decoction of the mashed berries and water was used to bathe the eyes.

Warning: All parts of *Phoradendron* are toxic and should be avoided.

CALTROP FAMILY (Zygophyllaceae)

The Caltrop Family is comprised of herbs and shrubs, and rarely, trees that exhibit a xerophytic or halophytic life history. The leaves are usually opposite and pinnately compound. There are approximately 30 genera and 250 species distributed primarily in tropical and subtropical areas. Six genera are native to the United States. Guaiacum wood and lignum vitae (a valuable heavy hardwood and wood resin used in some chemical tests) are the important economic products in this family.

Goathead, Puncture Vine (*Tribulus terrestris*)

Description: This is an introduced annual plant with trailing, hairy stems, and an extensive root system. The leaves are opposite and pinnate with four to eight pairs of leaflets. The flowers are yellow and are borne singly in the leaf axils. The fruit is hard, consisting of five spiny nutlets or burrs that break apart into five "tack-like" sections upon maturity. These burrs may injure livestock and are the bane of bicyclists. The plant is found in disturbed and waste places at the lower elevations.

Ecology & Ethnobotany: Puncture Vine has a 5,000-year history of medicinal uses, particularly in China and India. It was used for boosting the hormone production in men and women, and for urinary tract problems, itchy skin, and blood purification. The stems of the plant are considered to be astringent and act upon the mucous membrane of the urinary tract. A tea from the aboveground part of the plant is said to be good for arthritis.

Moore (1979) indicates that studies have been conducted on the seeds that show they are useful in the early treatment of elevated blood fats and cholesterol. It apparently helps prevent or lessen the severity of arteriosclerosis and atherosclerosis. Moreover, he says that the plant is useful in treating mild hypertension, contributing to a slower, stronger, more well-defined heart function, with greater relaxation between contractions and a lowering of the diastolic pressure.

FLOWERING PLANTS: MONOCOTS

Monocots are distinguished by a number of features. In monocots, the embryos of the seeds have only one cotyledon (seed-leaf). The leaves usually have parallel veins. The vascular bundles of the stems are irregularly arranged, and the cambium is lacking. The flower parts are arranged in threes or sixes, never in fives (or fours). Monocots are usually herbs, rarely shrubby.

ARROWHEAD FAMILY (Alismataceae)

Thirteen genera and 90 species in this family are found worldwide, of which five genera are native to the United States. Members of this family are aquatic and marsh perennial herbs. Their long-stalked leaves and even longer flowering stems are well adapted to wetland habitats. The flowers are comprised of 3 green sepals, 3 white petals, numerous stamens, and many pistils. This family is of little economic importance.

Caution: Aquatic plants such as Arrowhead and Water Plantain are sometimes found growing in polluted or contaminated water.

Water Plantain (*Alisma plantago-aquatica*)

Description: This is a perennial plant from fleshy, bulb-like stems. The basal leaves are long stalked and egg-shaped, and the flowers are white. Water

Plantain is usually found in marshes and ponds at lower elevations from Ventura County northward.

Ecology & Ethnobotany: he starchy, bulbous bases of Water Plantain are edible as a starchy vegetable (potato) after drying. Drying is said to remove the strong flavor. *Alisma* has a long history of use in Chinese medicine and is mentioned in texts dating back to about 200 A.D. It was also used by early herbalists as a diuretic and by the Cherokee Indians for application to sores, wounds, and bruises. It is described as a sweet, cooling herb that lowers blood pressure, cholesterol and blood sugar levels. The root was also used as a diuretic in the treatment of dysuria, edema, distention, diarrhea, and other ailments (Foster and Duke 1990). Water Plantain also furnishes food for water birds and muskrats.

Wapato, Tule-potato (*Sagittaria*)

Description: These are aquatic perennials with glabrous stems that grows 8-48 inches high and contains a milky juice. The leaves are large, arrow shaped, with the basal lobes, spreading outward. The stem is mostly leafless and is often branched. The flowers occur in whorls of 3, and are located near the ends of the stem. The sepals are ovate, whereas the petals are white. Tule potato grows in marshy places along the edge of ponds or streams or in wet meadows below 7,000 feet. Flowers in July and August. *Sagitta* is Latin for arrow, referring to the shape of the leaves.

"Quick Key" to the Tule-potatoes

1. Beak of mature achene less than 1/16-inch long and erect - *S. cuneata*
1. Beak of mature achene 1/8-inch long or more and horizontal - *S. latifolia*

Tule Potato (*S. cuneata*) This perennial occurs in ponds, slow streams, and ditches below 7,500 feet in the San Bernardino mountains.

Wapato (*S. latifolia*) This perennial occurs in ponds, slow streams, and ditches below 5,500 feet in the southern California mountains.

Ecology & Ethnobotany: All species produce starchy, white tubers that can be roasted or boiled, and then eaten. An important source of carbohydrates, the tubers contain a milky juice with a bitter flavor that is destroyed by heat. In the Pacific Northwest and

northern Rockies, the Lewis and Clark Expedition is said to have utilized these tubers extensively while exploring the Columbia River region. To many Native American tribes, Arrowhead was a primary vegetable. The small tubers are located at the ends of the long underwater rhizomes, perhaps a meter or more from the plant. They can be carefully removed without pulling up the whole plant. To collect the tubers, you can use your hands, a forked stick, or if the water is deep, your feet. Wade into the pond where the plants are growing and feel around for the tubers in the mud with your toes. They should feel like round lumps varying in size from peanut to potato. After dislodging them, the tubers usually float to the surface for easy harvesting. They are best developed in late summer or autumn. Boil or bake them like a potato to remove the poisonous properties, then peel and eat them. They can also be dried for future use. The tubers were used in a tea for indigestion, and in a poultice for wounds and sores. The leaves and stem contain alkaloids and may be poisonous.

SEDGE FAMILY (Cyperaceae)

There are 90 genera and 4,000 species worldwide, particularly in the cool temperate and sub-arctic regions. Twenty-four genera are native to the United States. In southern California, sedges are a large family of grass-like plants often found growing in wet places. Sedges make up a large proportion of the plants found at higher elevations. Most of the species have triangular stalks with the flowers arranged in spikelets. The family is of little economic importance, although several members are edible and medicinal. Sedge family members, however, should not be used medicinally without medical supervision since their properties and effectiveness are variable.

Sedge (*Carex*)

Description: Numerous species within this genus occur in southern California. They are normally difficult to identify in the field. A hand lens and mature fruit are often necessary for positive identification of most species. In general, they are perennial grasslike herbs with creeping rhizomes, short rootstalks, or fibrous roots, have 3-sided stems, and the leaves are 3-ranked. Sedges occur in a great variety of habitats from the foothills to the alpine zone.

Ecology & Ethnobotany: The young shoots and tender leaf bases of almost all species are sweet and furnish a tasty nibble. The fruits are also edible. *Carex* is a widespread genus with more than 1,000 species, making it available as an important emergency food in many parts of the country.

Sedge roots were used extensively as basketry material, particularly in coiled baskets. Some Native Americans would spend considerable time untangling roots in order to remove them in one piece. In fact, some areas were cleared of other plants to allow Sedge to grow long and untangled.

Nut-grass (*Cyperus*)

Description: Members of this genus are annual or perennial and grass-like. Leaves are mainly basal and three- ranked, and the stems triangular in cross-section. The spikelets are clustered in ball-shaped heads and the inflorescence consists of numerous heads borne on stalks radiating from the top of the stem and subtended by long, leaf-like bracts. The various species occur in wet, open soils along river banks, and the margins of lakes and ponds.

Ecology & Ethnobotany: The rhizome of *C. esculentus* (Chufa Flatsedge) bears small, nut-like, underground tubers that are edible. These may be eaten raw, boiled, dried and ground into flour, or roasted to a dark brown and ground into coffee. The species was so valued in ancient times that its tubers were placed in Egyptian tombs dating back to more than 2,000 years B.C. Other species may also have tubers or tuber-like structures. Volatile oils and astringent substances are found in a number of species that are used in perfumery and as remedies for digestive problems.

Nut-grass Drink

This is a Spanish recipe for a refreshing drink. Soak about ½ pound of tubers in water for 48 hours. Then mash the tubers, add 1 quart of water and 1/3 pound of sugar. Then strain the liquid through a sieve and serve as a drink (Fernald and Kinsey 1958).

Bulrush (*Scirpus*)

Description: The many species of Bulrush in California are perennials with grasslike or scale-like leaves. The various species can be found in marshy areas around lakes and ponds, and in other moist or wet areas from low to high elevations.

Ecology & Ethnobotany: The edible rhizomes of all species are quite starchy. hey may be eaten raw or baked, dried, or ground into a flour. As Olsen (1990) describes:

"...the young shoots just protruding from the mud are a delicacy raw or cooked. Furthermore, one can harvest them by wading into the water and feeling down along the plant until you come across the last shoot in a string of shoots that protrudes above the water. You then push your hands into the mud until the lateral rootstalk is encountered. By feeling along the rootstalk in a direction leading away from the last shoot, one can find a protruding bulb from which the new shoot is starting. This is easily snapped off and is edible on the spot."

The young roots when crushed and boiled also yield a sweet syrup. The pollen may be gathered and pressed into cakes and baked. The seeds may be used whole, parched, ground, or in mush. The stem bases may be eaten raw and are good for quenching thirst.

Scirpus stems, roots, and leaves can be used as the foundation and twining material in twined baskets. They were used extensively by Native Americans for making cordage, sandals, baskets, and mats. Waterproof "water bottles" can be made from the baskets by coating the inside with asphaltum. The stems were used for sleeping mats, padding, thatching dwellings, skirts, and sandals. Duck-shaped decoys were also made when hunting.

Leaves of *S. acutus* (also known as *Schoenoplectus acutus*) were used to poultice wounds and burns (Strike 1994).

IRIS FAMILY (Iridaceae)

There are about 70 genera and 1,500 species in this family worldwide, with the chief center of distribution in South America and tropical America. Five genera are native to the United States. In southern California, members of this family are perennials with narrow, grasslike leaves and thick rhizomes or fibrous roots. The flowers have three petals and three petal-like sepals joined on top of the ovary. The family is of economic importance as a source of ornamentals and saffron dye.

Iris (*Iris*)

Description: These perennials arise from a creeping rhizome. The stems are erect and the leaves are sword-shaped or linear. The flowers are large, occurring singly or in panicles. The perianth is comprised of 6 clawed segments that are united below into a tube. The outer three segments are spreading or reflexed whereas the inner ones are smaller and erect. Iris is the Greek word for rainbow.

"Quick Key" to the Irises

1. Flowers white to lilac with purple veins; plants of moist places - *I. missouriensis*
1. Flowers purple to blue-violet; plant of dry places - *I. hartwegii*

Wild Iris (*I. hartwegii*) This perennial with slender stems grows in dry woods from 5,000 to 7,500 feet. Flowers in May and June.

Western Blue Flag (*I missouriensis*) This slender perennial grows in moist areas as in meadows from 3,000 to 11,000 feet. Flowers in May and June.

Ecology & Ethnobotany: Members of the genus Iris contain irisin, an acrid resin concentrated mainly in the rhizomes, and present in the foliage and flowers. People who raise Irises sometimes develop a skin rash from handling the rhizomes. Cattle have died as a result of eating relatively large quantities of the plants. The rootstock produces a burning sensation when chewed. If eaten in quantity, Irises will cause diarrhea and vomiting. The poisonous rootstalks were used by Native Americans in a mixture of bile to poison arrowpoints.

Blue-eyed Grass (*Sisyrinchium*)

Description: The three species are perennial herbs with generally tufted, narrow leaves. The flowers have three petals and three sepals that are alike, pinkish-purple to blue in color. They are found in meadows and are often inconspicuous because of their tufted, grass-like leaves. The flowers open only in bright sunshine.

"Quick Key" to the Blue-eyed Grasses

1. Flowers orange to yellow - *S. elmeri*
1. Flowers blue, violet, or purplish - **2**

2. Leaves all basal with 1 leaf-like spathe subtending the flowers - *S. idahoensis*
2. Leaves present on main stem; 2 or 4 peduncles in axils of the leaves - *S. bellum*

    California Blue-eyed Grass (*S. bellum*) This tufted perennial grows in wet meadows in the yellow pine forest. Flowers from May to July.

    Yellow-eyed Grass (*S. elmeri*) This slender perennial grows in boggy and wet places from 4,000 to 8,500 feet. Flowers from July to August.

    Idaho Blue-eyed Grass (*S. idahoensis*) This perennial has a mostly leafless stem and grows in wet meadows and in moist areas from 4,500 to 10,800 feet. Flowers in July and August.

    Ecology & Ethnobotany: While the uses of most species is unknown, *S. bellum* was known among the Spanish-Californians as "azulea" and "villela." It was made into a tea considered to be a valuable remedy in treating fevers. It was thought that a patient could subsist for many days upon it alone.

RUSH FAMILY (Juncaceae)

    Nine genera and 400 species of Rushes are found in damp and wet sites of the cool temperate and subarctic regions. Two genera are native to the United States. These are grass-like annual and perennial plants with solid, rounded or flattened stems of wet and damp sites. The leaves are basal or alternate, and may be flat, folded, or round, and taper to a point. The flowers are small and have 6 undifferentiated sepals and petals (often termed tepals), 3-6 stamens, and a three parted ovary with many seeds. The family is of no direct economic importance to humans, although a few are ornamentals.

Rush (*Juncus*)

Description, Ecology & Ethnobotany: Many Rush species occur in southern California. In general, they are annual or perennial herbs often found in water or wet places. The flowers are in heads or panicles, and the tough, fibrous stems are inedible. However, they are useful in weaving baskets and mats.

Basketry

Baskets were an integral part of many Native American lives. They touched every aspect of their lives and were used in cooking and serving food, storage of goods and water, harvesting and winnowing seeds, and trapping fish. Weaving baskets demanded skill and patience.

There are two principle techniques used in basket weaving - coiling and twinning. In coiling, a foundation material comprised of a single stem or bundle of grass stalks was coiled around itself. Each coil was then bound and fastened to the one below it by stitching with a second type of pliable element, the wrapping. An awl made from bone or a spine from a cactus was used to make an opening in the lower coil through which the wrapping material was passed. Coiled baskets were often waterproofed by smearing hot pine pitch in the inside and outside for storing water.

In the twinning techniques, a fairly rigid material was the foundation around which two of three pliable fibers were interlaced. The start of a twinned basket looked like a starburst with the foundation sticks radiating out from a common center. As the twinning progressed, the sticks would be bent upward and pulled together to form the final shape of the basket. Twinning was used to make coarse, undecorated work baskets and to make openwork baskets such as winnowers and seed-beaters.

ARROWGRASS FAMILY (Juncaginaceae)

There are four genera and 26 species in the Arrowgrass Family. Two genera, including *Triglochin*, occur in the United States. The family is of no economic importance.

Arrowgrass (*Triglochin maritima*)

Description: This is a slender grasslike plant with fleshy basal leaves. Arrowgrass occurs in mountain swamps and around lakes below 7,500 feet in the San Bernardino mountains. Flowers April to August.

Ecology & Ethnobotany: The seeds of Arrowgrass can be parched and ground into flour. Roasted, they can be used as a coffee substitute. Seeds need to be parched since they contain cyanogenetic toxins that have caused death in livestock (Muenscher 1939). Parching or roasting the seeds renders them safe since the poison is volatile (Kirk 1975, Harrington 1963).

The young white leaf bases were collected around April or May from the inner leaves of the basal cluster. These leaf bases, when eaten raw at the right stage have a mild, sweet cucumber-like taste. They are generally better if cooked. In springtime, the leaf bases contain few toxic compounds, whereas the mature leaves and flower stalks should never be eaten. The leaves contain hydrocyanic acid, a toxin that interferes with the uptake of oxygen. Symptoms include headache, heart palpations, dizziness, and convulsions.

Caution: These plants are toxic when fresh. Several references list these plants as livestock poisoners until they dry and then cyanogenic properties evaporate, breakdown or dissipate.

DUCKWEED FAMILY (Lemnaceae)

Plants in this family are small and float on slow or stagnant waters. They have small thread-like root hairs that obtain nutrients from the water. All Duckweeds are used as food by wildlife and have been recorded from the stomachs of ducks. Two genera can be encountered: *Lemna* and *Spirodela*.

"Quick Key" to the Duckweeds

1. Roots solitary on each plant - **Lemna**
1. Roots usually 2 or more on each plant - **Spirodella**

Duckweed (*Lemna*)

Description: These are small plants, often not much larger than a pinhead.

Ecology & Ethnobotany: Under survival conditions, Duckweed can provide copious and palatable material for salads. Additionally, the Maidu and Miwok Indians in California used Duckweed as a diuretic and a general tonic.

Duckmeat (*Spirodella polyrhiza*)

Description, Ecology, & Ethnobotany: This species is frequently associated with *Lemna*. Duckmeat is a coarse species with a purplish-tinged lower side. As with *Lemna*, Duckmeat can also provide copious and palatable material for salads. The Chinese use this species to treat hypothermia, flatulence, and acute kidney infections.

Aquatic Plants and Water

The days of safely drinking water straight out of the mountain stream are long gone. That also goes for eating aquatic plants such as Cattail (Typha), Duckweed (Lemna & Spirodela), Watercress (Rorippa), Pond Lily (Nuphar) and Bulrush (Scirpus). Even in the most remote areas of the country, there are a number of disease-causing bacteria, protozoans, and viruses. A fairly common intestinal disorder among backcountry hikers is caused by the protozoan Giardia lamblia, which is carried in the intestines and feces of muskrats, beavers, moose, voles, and other water-loving mammals.

Other intestinal diseases also can be found in what appears to be clean water sources. Cases of Campylobactor, E. coli, and type A hepatitis have been traced back to drinking untreated water in the backcountry. Additionally, cities, farms, and suburbs can experience clean water problems resulting from oil spills, pesticides, and toxic pollutants from mining operations.

In areas where edible plants are found, it is best to treat the water and the plants that grow in it. One method would be to soak the fresh greens in a disinfectant if pollution is suspected. You can use any of the water purification tablets on the market, or a teaspoon to tablespoon of chlorox (bleach) in a quart of water. Then rinse the plants well and prepare. Obviously, if the waters are polluted with oils, pesticides, or other toxic pollutants, this method will not work. In any case, extreme care should be exercised.

LILY FAMILY (Liliaceae)

This family contains many beautiful wildflowers. It is a large and varied family with approximately 250 genera and 4,000-6,000 species worldwide. About 75 genera are native to the United States. The family is characterized by a perianth of six parts, with a superior ovary and a 3-lobed stigma. The fruit is a capsule that splits open when ripe. The family is a source of many ornamentals, several important fibers, fermented and distilled beverages, and steroidal compounds. While some members of this family are edible, there are many poisonous species. For example, you can boil and eat the bulbs of the True Lilies (*Lilium*), but some genera contain highly toxic alkaloids. Nearly two hundred alkaloids and numerous glycosides occur in the family.

Onion (*Allium*)

Description: The many species of Allium arise from bulbs, and all have the characteristically distinct onion odor. The odor is apparently caused by the presence of volatile sulphur compounds in all parts of the plant (causing their strong flavor and irritation to eyes). In all, there are about 300 species of onions in the world. Some other common names of Allium include leeks, garlic, and chives. The small flowers are clustered together in umbels. Onions are found in a variety of habitats from low elevations to the alpine zone. Allium is the ancient name for garlic. The derivation of this name may be from the Celtic all, which means "pungent."

"Quick Key" to the Onions

1. Leafless flowering stem $\frac{3}{4}$- 6 inches tall; leaves longer than scape - **2**
1. Leafless flowering stem 6-20 inches tall; leaves shorter than scape – **4**

2. Leaves 2 - *A. tribracteatum*
2. Leaves one - **3**

3. Leaf rounded in cross-section; stamens shorter than petals; petals lanceolate - *A. fimbriatum*
3. Leaf flat; stamens longer than petals; petals linear - *A. burlewii*

4. Flowers white to pale pink; leaves becoming rolled up and filiform - *A. amplectens*
4. Flowers pale rose; leaves flat - *A. campanulatum*

Narrow-leaved Onion (*A. amplectens*) This herb grows in the Cuyamaca Mountains below 6,000 feet and also further north. Flowers from March to June.

Burlew's Onion (*A. burlewii*) This perennial grows on dry slopes and ridges from 6,000 to 9,000 feet in the San Jacinto Mountains and northward. Blooms from May to July.

Sierra Onion (*A. campanulatum*) This is a herb that grows on dry slopes and in the woods between 2,000 to 8,500 feet. Flowers from May to July.

Fringed Onion (*A. fimbriatum*) This herb grows on dry slopes and flat areas from 2,000 to 8,000 feet. Flowers from March to July.

Three-bracted Onion (*A. tribracteatum*) This perennial grows in the San Jacinto Mountains and north at 4,000 to 8,000 feet. Flowers from April through July.

Ecology & Ethnobotany: All *Allium* species are known to be edible. The bulbs may be eaten raw, boiled, steamed, creamed, in soup, and are especially good when used as a seasoning. Ingestion of large amounts of Onions, including the cultivated ones, can cause poisoning or cause goiter, but are otherwise not known to be harmful. Regardless, eating them in moderation is the key. The plants are valuable in all seasons and can be used as greens and as flavoring. The seeds and leaves can also be eaten. Onions will keep a long time, because the skin dries and preserve the flesh inside. Wild Onions do contain large amounts of some important micronutrients, more Vitamin C than an equal weight of oranges, and more than twice as much Vitamin A as an equal weight of spinach (Kindsher 1987). Additionally, onions contain a significant amount of a starch called inulin, which is not easily digested by humans.

Medicinally, Onions have a number of uses. Soldiers during World War I took advantage of their natural antiseptic properties by applying Allium juice to wounds to prevent infection. The juice of Wild Onions can be boiled down until it is thick and used as a treatment for colds and throat irritations. The juice was also used as an insect repellent when rubbed over the body. The onion smell apparently has some beneficial effects on the circulatory, digestive, and respiratory systems.

Warning: Wild Onions should not be confused with the so-called "poison onions" or Death Camas. *Zigadenus venenosus* is most likely to be confused with *Allium*. These are bulb-bearing plants with grasslike leaves, also in the Lily Family. They have upright, more elongated (not umbrella-like) clusters of white or cream flowers. They contain highly toxic alkaloids and all plant parts, including the bulbs, can be fatal if ingested in any quantity. They also lack the characteristic strong odor of onions.

Brodiaea (*Brodiaea terrestris*)

Description: The flowering stems are erect and the leaves are few, basal, and grasslike. The leaves often wither away before the flowers appear. The flowers occur in an umbel and the segments all look alike. The fruit is a capsule. This species occurs on slopes and flats below 5,000 feet from Kern County to San Diego County. Flowers April to June.

Ecology & Ethnobotany: The corms of most Brodiaea species are edible raw, but are somewhat muscilagenous. It is better if they are boiled for a few minutes or roasted. They can also be mashed and dried for future use in stews. Since the corms grow deep, it is usually easiest to harvest them with digging sticks.

The crushed corms were used as a paste which was smoothed over the sinew backing on bows. The paste was also used to bind paint pigments to hunting bows.

Brodiaea corms and flowers were used as soap and shampoo.

Caution: When collecting bulbs, be aware that the poisonous *Zigadenas* (Death Camas) may be in the area too.

Digging Sticks

Roots and bulbs of many plants are best collected with the aid of a digging stick. A digging stick is about 2-3 feet long, 1-2 inches thick, beveled at one end and fire hardened. To use, thrust the stick into the ground beside the plant and pry upward while pulling on the plant from above.
To fire harden your digging stick, simply hold the point a few inches above a bed of hot coals and slowly turn it as you would a skewer. Take care not to char the wood, but let it turn to a light brown color.

Mariposa Lily, Sego Lily (*Calochortus*)

Description: They are characterized as perennials from bulbs, with tulip-like flowers that are few and showy. Mariposa is the Spanish name for butterfly. These species can be found in dry open places from low to mid elevations.

"Quick Key" to the Mariposa Lilies

1. Flowers yellow - *C. concolor*
1. Flowers not yellow - **2**

2. Flowers red to orange - *C. kennedyi*
2. Flowers white, lavender, to blue or purple - **3**

3. Flowers deep lilac; petals $1\frac{1}{2}$ - 2 inches long - *C. splendens*
3. Flowers white to dull lavender; petals 3/4 to $1\frac{1}{2}$ inches long - **4**

4. Bracts 3/8 to $\frac{3}{4}$-inch long; flowers not in umbels; plant of moist area - *C. palmeri*
4. Bracts $\frac{3}{4}$ - 2 inches long; flowers in an umbel of 2 - 4 flowers; plant of dry area - *C. invenustus*

   Goldenbowl Mariposa (*C. concolor*) This is an erect perennial with slightly branched stem that grows on dry slopes and in gravelly areas from 2,000 to 7,500 feet. Flowers from May to July.
   Plain Mariposa Lily (*C. invenustus*) This is a perennial plant with slender, simple and erect stems grows on dry soils from 4,500 to 9,000 feet. Flowers from May to August.
   Red Mariposa Lily (*C. kennedyi*) This erect perennial grows on dry, rocky slopes from 2,000 to 6,500 feet in desert areas in Kern, Inyo, Ventura, and San Bernardino Counties. Flowers from April through June.
   Palmer's Mariposa (*C. palmeri*) This erect perennial grows in moist meadows from 3,500 to

6,500 feet in the San Bernardino Mountains and northward. Flowers from May to July.

Splendid Mariposa Lily (*C. splendens*) This perennial with erect, branched stems grows on dry slopes up to 8,500 feet. Flowers in May and June.

Ecology & Ethnobotany: While the entire *Calochortus* plant is edible and can be used as a potherb, the highly nutritious bulbs are usually sought. They are smaller than walnuts and may be eaten raw, boiled or roasted in hot ashes in pits, or steamed before eating. The bulbs are dug in the early spring, usually before flowering. The bulbs can also be threaded on a string and dried with or without cooking first. They can also be dried and ground into flour or cooked and mashed into cakes for preservation. *Calochortus* bulbs were eaten by many Native Americans and were  considered an important food source. When numerous bulbs were collected, they were usually pit-cooked. The flower buds can be eaten and have a sweet taste. Seeds are also edible.

However, care should be taken not to over harvest these plants. This is particularly true in Utah where *C. nuttallii* (Sego Lily) is the State Flower.

Note: These plants are becoming increasingly rare in some areas, primarily due to habitat destruction and overgrazing. Harvesting the corms destroys the plants. Because of the plant's rarity and beauty, their use today is not recommended.

Soap Plant, Amole (*Chlorogalium pomeridianum*)

Description: The stems of this plant grow 2-5 feet and are nearly leafless. There are many basal leaves that are keeled and wavy-edged. The white flowers occur in a long spreading cluster and are purple veined. The bulb which can grow to about 2 inches in diameter has a dense coat of coarse fibers. The species is found in dry, open and stony ground up to 5,500 feet. Flowers May to August.

Ecology & Ethnobotany: The delicate flowers open only in the afternoon. The bulbs can be eaten after being cooked in an earth oven. The young shoots can be eaten in the spring.

The baked bulbs were used as a poultice on skin sores. The scales of the bulb form a lather with water. Fish can be caught by throwing crushed Amole bulbs into a stream that has been dammed. This stupefies the fish causing them to float to the surface to be collected.

Wild Hyacinth (*Dichelostemma capitatum*)

Description: This is a perennial herb with a slender scape up to 2 feet tall. The blue-violet flowers occur in a dense, head-like cluster of 4-10 flowers. The cluster is subtended by 4 ovate purplish bracts. This is a common plant on dry hills and plains. Flowers March to May.

Ecology & Ethnobotany: The flowers and bulbs are edible raw or cooked.

Fritillary (*Fritillaria pinetorum*)

Description: This erect herb has a simple, glabrous stem that is 4-12 inches tall. The leaves are linear and number 12-20. The terminal flowers are erect or slightly nodding, purplish, and mottled with greenish yellow. The perianth segments are all alike. Fritillary grows on shaded, granitic slopes from 6,000 to 10,500 feet. Flowers from May to July. The genus name comes from the Latin *fritillus*, meaning "dice box," in reference to the short, broad capsule characteristic of the genus.

Ecology & Ethnobotany: The bulbs of this genus have been a staple for Native Americans since prehistoric times. Bulbs of all species are edible raw or cooked, but are relatively rare and should be considered only in an emergency.

Lily (*Lilium*)

Description: The true Lilies have large showy flowers that are usually spotted. The flowers may be funnel-shaped, or the tepals maybe curved backward. The color varies from yellow to orange, with one to several flowers appearing at the top of a single stalk up to 3 feet tall. Leaves are alternate or whorled. The roots are covered with scales, an important identifying characteristic.

Leopard Lily (*L. paradalinum*) This stout perennial grows in large colonies along stream banks and in springy places below 6,000 feet. Flowers from May through July.

Lemon Lily (*L. parryi*) This slender perennial grows in wet, springy areas, moist meadows, or along wet banks from 4,000 to 9,000 feet. Flowers in July and August.

Ecology & Ethnobotany: The bulbs of all members of this genus are edible. They can be eaten raw or steamed, and have a bitter or peppery taste. Some suggest that the plants are better used as a flavoring agent and that the bulbs taste better after flowering. One way to locate bulbs in the fall is to flag the plant(s) during the summer. Because of their relative rarity and beauty, they should not be harvested except in an emergency.

Nolina (*Nolina parryi*)

Description: This perennial plant resembles a Yucca. It has a thick woody base or short trunk that is 8 inches to 6 feet tall. The leaves are gray-green and numerous, and occur in a dense crown at the top of the trunk. Margins of the leaves are finely toothed, not as stiff and rigid as in yuccas. The numerous flowers are small, and form a large, terminal, flowering cluster on a stalk that is 3 to 4 feet long and arises above the leaves. The individual flowers are $\frac{1}{4}$-inch long, and cream to white in color. Nolina grows on the dry slopes of mountains up to 5,500 feet bordering the deserts. Flowers in May and June. This species was once included in the Agave Family (Agavaceae).

Ecology & Ethnobotany: The fruit and stalks can be used as a food. Fiber used to make rope and cordage.

False Solomon's-seal (*Maianthemum*)

Description: These are annual herbs with extensive, horizontal rootstalks. Leaves are alternate and sessile or on short petioles.

"Quick Key" to the False Solomon's-seal

1. Perianth segments less than 1/8-inch long; fruit a red berry - **M. racemosum**
1. Perianth segments $\frac{1}{4}$-inch long; fruit a purple to black berry - **M. stellatum**

Feathery False Solomon's-seal (M. racemosum) This is an erect perennial that grows from 5,000 to 8,000 feet. Flowers in June and July.

Starry False Solomon's-seal (M. stellatum) This erect perennial grows in wet or brushy places from 4,000 to 8,000 feet. Flowers from April through June.

Ecology & Ethnobotany: Both species have edible berries that are not especially palatable. If eaten in quantity they can act as a laxative. Cooking the berries removes much of the purgative elements making them a bit more palatable. They are also high in Vitamin C. The young shoots and leaves can be used like asparagus or eaten as a potherb.

False Solomon's-seal have starchy rootstocks that may be eaten. However, the rootstocks must be soaked overnight in lye. The Ojibwa Indians of Ontario, Canada, used the white ashes from their fire pits instead of lye, which supposedly removed the bitterness. The roots are then boiled and rinsed several times to remove the lye. A tea made from the roots was used for headaches.

A tea from Starry False Solomon's-seal was used by some Native Americans as a contraceptive. The powdered roots were used on

wounds to stop bleeding, and a root decoction was used internally as a tonic or externally as an antiseptic wash for infected sores or wounds. The mashed root of Starry False Solomon's-seal was thrown into a stream as a fish stupifier, making the fish easier to catch. A decoction of Feathery False Solomon's-seal was used as a contraceptive, to regulate menstrual disorders, to relieve kidney problems, to heal wounds and as a heart tonic.

Corn Lily (*Veratrum californicum*)

Description: This is a tall, stout perennial that has leafy stems and grows 3 to 6 feet tall. The leaves are broad and clasping, strongly veined, and ovate to elliptic in shape. The cream to white flowers occurs in a dense, terminal, branched cluster. The 6 perianth segments are all alike, and with a green gland near the base. Corn Lily is common in wet meadows and along stream banks, particularly at higher elevations. Usually found below 11,000 feet. Flowers in July and August.

Ecology & Ethnobotany: These plants are very poisonous if ingested and have an inconsistent mixture of several powerful alkaloids. Some of the symptoms include depressed heart action, salivation, headache, burning sensation in the mouth, slowing of respiration, and death from asphyxia. These violent symptoms of poisoning may occur within 10 minutes. Avoid any use of the plant that involves ingestion. In some cases, just handling Veratrum can cause severe itchiness and irritation. The water in which the roots of both California False Hellebore was

 boiled is considered effective in killing head lice. The powdered roots have also been used as an insecticide. Even nectar in the flowers is poisonous to insects and can cause serious losses among honeybees.

Yucca, Spanish Dagger (*Yucca schidegera*)

Description: This tree occurs on dry, rocky slopes in the Mojave Desert and the southern desert areas, mainly below 5,000 feet. Flowers from April to May. Yuccas were once included in the Agave Family (Agavaceae).

Ecology & Ethnobotany There is a unique relationship between Yuccas and the Yucca moth (*Tegeticula maculata*). This is the only pollinator for the plant. The mouthparts of the female moth are such that it gathers the pollen and forms it into a ball which is then scraped across the next stigma. In addition, the moth lays its eggs only in the seedpod of the Yucca. Without each other, both of these organisms could not exist.

All species of Yucca provided a plentiful and dependable food source for many Native Californians. The large pulpy fruits were eaten raw or roasted, or cooked and formed into cakes or dried for future use. The flowers buds were also roasted and eaten, and they have a high sugar content which served as a sweet treat for the children. A tea can be made from the seeds or the seeds were pounded and cooked as mush.

The leaves can be pounded to release the fibers for use in basketry and cordage making. One method used by the Diegueno was to bury the leaves until the fleshy part rotted away, yielding a fine white fiber. The tough fibers were also used in making nets, hats, sandals, and mattresses.

The chopped or pounded roots yield a soap-like substance. It can be used for general washing. Paintbrushes can be made by fringing the smaller leaves.

Death Camas (*Zigadenus venenosus*)

Description: This is a perennial plant that grows from an oblong, ovoid bulb. The stem is simple, glabrous, and 10 to 24 inches long. The leaves are glabrous and linear, with the basal leaves folded and with rough margins. The main leaves have scarious sheaths. Flowers occur in a terminal raceme that is 2 to 8 inches long. The white perianth segments are all alike, clawed or narrowed at the base and with toothed, upper margins. Death Camas

grows in moist, grassy areas below 8,200 feet. Flowers from May through July.

Ecology & Ethnobotany: Death Camas is very poisonous plants if ingested. The alkaloids, primarily concentrated in the bulbs, can cause muscular weakness, slow heartbeat, subnormal temperature, stomach upset with pain, vomiting, and diarrhea, and excessive watering of the mouth. Death Camas should not be confused with plants such as the edible Camas (*Camassia*), which formed a staple food for aboriginal peoples in the Northwest. It is also difficult to distinguish Death Camas from other edible plants, including Wild Onion (Allium), Sego Lilies (*Calochortus*), Fritillaries (*Fritillaria*), and Brodiaeas (*Brodiaea*) prior to flowering.

Crushed Death Camas bulbs were used by some Native Americans as poultices for boils, bruises, strains, rheumatism, and in some cases rattlesnake bites.

ORCHID FAMILY (Orchidaceae)

This is one of the largest families in the world with the greatest concentration of species found in the tropics. There are over 15,000 species in several hundred genera. The flowers are irregular in shape with three sepals and three petals. One of the petals forms a lip, sac or pouch on the lower side of the flower. The flower structure is highly specialized for insect pollination. The family is an outstanding source of ornamentals, and although a few orchid species have utilitarian uses, the majority are rare and should only be considered for use in emergency situations.

Spotted Coralroot (*Corallorrhiza maculata*)

Description: They are typically yellowish to brownish-red perennials with coral-like rhizomes. *Corallorrhiza* means "coral root." The rhizomes are associated with fungi that aid in the uptake of nutrients. This species occurs in the coniferous forests between 5,000 and 9,500 feet from San Diego County northward. Blooms June to August.

Ecology & Ethnobotany: Consumption of the toxic rhizome can cause hyperthermia and profuse perspiration. The rhizome of has been used as a diaphoretic, febrifuge, and sedative, and the dried stems were used by the Paiute and Shoshone Indians of Nevada to make a tea to build up blood in pneumonia patients (Coffey 1993, Foster and Duke 1990). Strike (1994) indicates that the plant was used to reduce fevers or as a sedative.

Helleborine (*Epipactis gigantea*)

Description: Helleborine is a perennial with one to several erect stems arising from rhizomes. The stems are covered with numerous, sheathing leaves. The showy flowers form a raceme with all flowers on one side of the stem. The sepals are coppery-green with light brown venation. This species occurs along moist stream banks below 7,500 feet. Flowers May to June.

Ecology & Ethnobotany: The name Epipactis derives from a classical name used by Theophrastus (circa 350 B.C.) for a plant used to curdle milk. There are about 25 species found worldwide, mostly in Europe (15 species) and Asia. Two species are found in the U.S. - E. gigantea and E. helleborine. The latter is thought to have arrived in North America from Europe in the 19th century and grows in the northeastern U.S.

Native Americans made a decoction of the fleshy roots for internal use when they felt "sick all over."

Rein Orchid (*Platanthera*)

Description: These are perennials, often with fleshy or tuberous roots. The small white to yellowish-green flowers are in spike-like racemes. At the base of the lip is a spur. Rein Orchids can be found in forest understories, or in wet areas and meadows. Three species can be found in the southern California mountains.

"Quick Key" to the Rein Orchids

1. Leaves all basal, and are withered at flowering time; stem leaves are reduced to bracts - *P. elegans*
1. Leaves on main stem are reduced, but not bract like; leaves are not withered at flowering time - **2**

2. Flowers white; spike dense and many flowered - *P. leucostachys*
2. Flowers greenish; spike slender and few-flowered - *P. sparsiflora*

Rein Orchid (*P. leucostachys*) Rein Orchid grows in wet, springy, or boggy places below 11,000 feet. Blooms from May through August.

Elegant Piperia (*P. elegans = Piperia e.*) Elegant Piperia grows in dry woods below 8,000 feet. Flowers from May to September.

Few-flowered Bog Orchid (*P. sparsiflora*) Few-flowered Bog Orchid grows along streams and in boggy places, from 4,000 to 11,000 feet. Flowers from June through August.

Ecology & Ethnobotany: The tuber-like roots of many species may be eaten raw or cooked. However, it is recommended that anyone experimenting with these plants take a cautious approach until the poisonous nature of the plants is clarified. Some Native Americans used extracts from these plants as poison to sprinkle on baits for coyote and grizzly bear.

Hooded Ladies'- tresses (*Spiranthes porrifolia*)

Description: This is a small herb with fleshy roots. The white flowers are in a dense spirally twisted spike, with the sepals and lateral petals appearing to be fused and forming a hood around the column. This species is found in springy places below 8,000 feet in the San Bernardino Mountains. Flowers July to August.

Ecology & Ethnobotany: The species was named after Nikolei Rumliantzev, Count Romanzoff, a Russian patron of science.

According to Weedon (1996), *S. romanzoffiana* has strong diuretic properties, making it undesirable for eating. Additionally, Foster and Duke (1990) indicate that Native Americans used a plant tea of *S. cernua* (Nodding Ladies'-tresses) of eastern United States as a diuretic for urinary disorders, venereal disease, and as a wash to strengthen infants. They also state that other North American, European, and South American species have also been used as diuretics and aphrodisiacs.

GRASS FAMILY (Poaceae)

With 600 genera and 10,000 species of grasses worldwide, the Grass family is the most common family of flowering plants found in practically all habitats and on all continents. There are over 180 genera and nearly 1,000 species in the United States. Grasses have round, hollow stems with linear sheathing leaves. Because grasses are wind pollinated, they have no showy flowers to attract insects. The flowers have been reduced to scaly bracts that enclose the male and female parts. The grains form within the papery bracts after pollination. The grass family contains many of the most economically important plants in the world, including all the important cereal crops and forage grasses essential to raising

446

domesticated livestock. Grasses can be found from alpine meadows sea level.

Grains from several different species of plants were often mixed together to enhance the flavor of the pinole. All grasses in the southern California mountains have edible grains that are generally small and tedious to collect. The small seeds are tightly enclosed in scales which are hard to remove. The larger grains of some grass species were a staple among aboriginal peoples. The grains were harvested with a beater and ground for mush and flour. The grains are rich in protein and can be eaten raw, but are better if roasted, ground into flour, or boiled into mush. They may also be boiled in the same way as rice, and added to soups or stews. The reported toxicity of some Grass species may be that of a fungus (e.g., *Claviceps purpurea*) associated with the grasses. Any inflorescences containing black grains should be discarded since they may have a harmful fungus infection.

All bladed grasses are edible and are rich in vitamins and minerals. Animals often consume grasses to get nutrients they can't get elsewhere. The young shoots are edible raw, and are not as fibrous and therefore easier to digest than mature leaves. The green or dried leaves can be steeped to make a tea. Following are some genera of grasses used by Native Americans.

Note: Grasses that are infected with *Claviceps purpurea* develop purple sclerotia (ergots) in place of the healthy grain. The sclerotia contain a number of toxic alkaloids and if eaten, can cause severe illness and sometimes death. One effect of the toxins is constriction of the blood vessels, whereby the impaired circulation may result in gangrene or loss of limbs. Another effect is on the nervous system, which results in convulsions and hallucinations (Webster 1980).

The sclerotia of *C. pupurea* are used medicinally to hasten uterine contractions during childbirth. The ergot of commerce is produced by cultivating the fungus on Rye (*Secale cereale*) and other plants. Attempts are being made to extract the medically important alkaloids from pure cultures of the fungus (Webster 1980).

Wild Flours

Pinole -- was made using small grains which were parched by tossing in a basket with glowing coals or hot pebbles, keeping the grains in constant motion. Grains were then pulverized and eaten. Sometimes the pulverized grains were pressed into cakes held together by the grains' natural oil, with no other liquids added.

Ricegrass (*Oryzopsis hymenoides*)

Description: A perennial bunchgrass from 8 to 20 inches tall, Ricegrass is found in dry rocky or sandy ground, grasslands, and valleys from low to mid elevations.

Ecology & Ethnobotany: The relatively large grains of this species have been used by Native Americans as food for centuries. The grains were collected with the stems and held over a fire to singe off the fine white hairs. They can also be collected in a pan or basket with hot coals or rocks, and shaken to burn off the hairs. The grains were then ground into flour and used as mush, to thicken soup, or made into cakes.

Common Reed (*Phragmites communis*)

Description: Common Reed is native to every continent except Antarctica. It often forms dense thickets along streams, ditches, and marshes at low elevations. The record of human uses for this species is extensive. The reeds have been used for roof insulation since before the time of Christ. Because it is light, it is best known for its use in arrow shafts. The roots can be eaten raw or cooked.

Ecology & Ethnobotany: Meal can be obtained from pulverized stems. In the fall, the leaves and stems may become encrusted with grayish exudate. This exudate, actually honeydew (excreta of whitefly and aphids), was obtained from stalks. Stalks were cut and flayed to remove honeydew crystals, which were winnowed and cooked into stiff dough. Dough was formed into cakes, sun dried, and stored. Split culms provided fiber. Common Reed was used to make flutes and other musical instruments in addition to carrying nets and cordage. The honeydew was given to pneumonia patients to loosen phlegm and soothe pain in lungs.

PONDWEED FAMILY (Potamogetonaceae)

Of the two genera and 100 species worldwide, *Potamogeton* is native to the United States. The family is of no direct economic importance to humans, but provides a valuable source of food for ducks and other wildlife.

Pondweed (*Potamogeton*)

Description: Pondweeds are perennial aquatic plants with extensive, slender rhizomes and simple branched stems that often root at the nodes. The leaves have stipules that clasp the stem. The lower leaves are alternate and submerged, and the upper often floating leaves are wider and opposite. The small, greenish flowers

are clustered in a spike that arises from the upper leaf axils. Most Pondweed species can be identified without flowers and fruits, but the floating or submersed leaves may be necessary. Hybridization among species is fairly common. There are many species of *Potamogeton* in southern California.

Ecology & Ethnobotany: Probably all Pondweeds have starchy, edible rhizomes, but species with larger rootstocks are preferred for gathering. *Potamogeton diversifolius* (Waterthread Pondweed) was an important source of strong fibers which were rolled into cordage to make carrying nets, rabbit-trap nets, and other items (Strike 1994).

<u>BUR-REED FAMILY (Sparganiaceae)</u>

One genus with about 20 species occurs in this family. Bur-reeds are found primarily in the cooler regions of the north temperate zone, Australia, and New Zealand. Several species are native to the United States. The plants arise from creeping rootstalks and the roots are fibrous. The species have no direct economic importance to humans.

Bur-reed (*Sparganium*)

Description: These are aquatic perennials with unbranched, erect or floating stems. The leaves are linear and sheath the stem. The flowers are borne in dense, round clusters. Bur-reeds can be found in shallow waters of marshes, ponds, and slow-moving steams.

Ecology & Ethnobotany: The bulbous bases of the stems and tubers of *S. eurycarpum* (Broadfruit Burreed), *S. simplex* (Simplestem Burreed) and *S. angustifolium* (Narrowleaf Burreed) can be used as food in much the same way as cattails (*Typha*) and bulrushes (*Scirpus*): dried and pounded into flour.

CATTAIL FAMILY (Typhaceae)

There is one genus (*Typha*) with about 15 species worldwide in this family. Members are marsh or aquatic perennials with creeping rootstalks. Two kinds of flowers are borne in crowded spikes, with the staminate (male) flowers above and the pistillate (female) flowers below. The family is of little economic importance. This family now incorporates the family Sparganiaceae.

Cattail (*Typha*)

Description: The two species, *T. angustifolia* (Narrowleaf Cattail) and *T. latifolia* (Broadleaf Cattail), found in California are also found over much of North America. *Typha* is Greek for "cattail." It is often referred to as "Cossack asparagus." Cattails reproduce rapidly in marshy areas that are often unsuited for agriculture.

Ecology & Ethnobotany: In his book *Mountain Man*, Vardis Fisher (1965) illustrates how Indians utilized Cattails:

"And what be that?' asked Sam, staring at the mold. He knew that Indians ate just about everything in the plant world, except such poisons as toadstools, larkspurs, and water parsnips. It was a marvel what they did with the common cattail - from the spikes to the root, they ate most of it. The spikes they boiled in salt water, if they had salt; of the pollen they made flour; of the stalk's core they made kind of a pudding; and the bulb sprouts on the ends of the roots they peeled and simmered."

Virtually every part of the plant has a use, from food to fiber. Euell Gibbons considered the Cattail the "supermarket of the swamps." Although both Cattail species have edible rhizomes, the rhizomes should never be raw since they may cause vomiting. The rhizomes should be boiled or roasted, or dried and then ground into meal or flour. Another way to obtain the starch from the rhizomes is to follow a technique described by Euell Gibbons (1962):

"...after scrubbing the root [i.e., rhizome] and peeling off the spongy outer layer surrounding the white stiff core, cut the core into small sections, and place the pieces in a bowl of cold water. Work the core with your hands, separating the fibers and scraping out the starch. Slosh the fiber around in the water until you have removed all the starch. Pour off the water through a course sieve to extract the fibers. Allow the water to settle for a little while the starch settles to the bottom of the container. Then carefully pour off the water, leaving the starch in the bowl. For a cleaner starch, pour in some more water and let it settle again. Then pour off the water. After this, you can use the starch almost immediately to make pancakes, breads, and biscuits."

Harrington (1967) reports that one acre of Cattails can yield over three tons of nutritious flour, although extraction techniques need to be refined for commercial exploitation.

When pulling up the rhizome, you may notice newly emerging buds. These can be scrubbed, peeled, and eaten raw or boiled. The swollen joint between bud and rhizome is also starchy. Peel it, then roast or boil for a potato-like vegetable. Like the rhizomes, this part should not be eaten raw. The young green shoots can be peeled of their green outer layer and eaten raw or cooked. It is always good to boil them in a couple of changes of water if there is any bitterness. The peeled core can also be sliced and added to salads.

While the flower spikes are still green, remove the papery sheath and boil the cluster for a few minutes. The flower spikes can then be eaten like "corn on the cob", although the core of the cluster is inedible. Cattail pollen is high in protein, and can be used in flour for breads or eaten raw. However, if you are allergic to pollen, it should be avoided. The seeds from the female portion of the flower spike can be pulverized to make a nutritious, protein rich flour. Seeds can be extracted from the fluff by parching them.

Useful fibers can be derived from Cattails. Fibers in stems can be loosened by soaking plant material in water for several days. The silky fluff on the seeds is buoyant and water repellent and makes a good insulator, especially in boots. The silk can be used for stuffing items from pillows to down vests. It can also be used for tinder. The fuzz will explode into flame with a spark from a flint and steel set. Leaves can be woven to make mats, sandals, baskets, etc. The stems provide a good coil foundation for baskets. Additionally, the stalks have been used as arrows and hand drills. A toothbrush can be fashioned from the fuzzy stem with the flowers removed.

Medicinally, the chopped or pounded rhizome was applied to the skin for minor wounds and burns. Cattail down was used as dressings for wounds. Brown (1983) indicates that a sticky juice derived from between the young leaves can be used as a styptic, antiseptic, and anesthetic. The jelly from between the young leaves can be applied to wounds, sores, external inflammations, and boils to soothe pain. Brown (1983) also indicates that the jelly was rubbed on the gums as a Novocain substitute for dental extraction.

Cattail Jelly

Jelly from Cattails? The following is a recipe from Jan Phillips, author of Wild Edibles of Missouri, *who says she makes jelly after the first flour has been rubbed out. The jelly is made by boiling the roots (rhizomes) for 10 minutes in enough water to cover them. For every cup of liquid, add equal amounts of sugar and a package of pectin per every four cups of juice. The jelly somewhat resembles honey in both taste and color.*

"We are losing our ancestral knowledge because the technicians only believe in modern science and cannot read the sky."

— Andean peasant expression

REFERENCES

Altschul, S. 1973. Drugs and Food from little-known Plants. Harvard University Press, Cambridge, Mass.

Anderson, J.P. 1939. Plants used by the Eskimo of the Northern Bering Sea and Arctic Regions of Alaska. American Journal of Botany 26(9):714-716.

Angier, B. 1974. Field Guide to Edible Wild Plants. Stackpole Books, Harrisburg, PA.

Angier, B. 1966. Free for the Eating. Stackpole Books. Harrisburg, PA

Bailey, F.L. 1940. Navajo Foods and Cooking Methods. American Anthropologist 42(2):270-290, April-June.

Balls, E.K. 1962. Early Uses of California Plants. Berkeley and Los Angeles, University of California Press.

Barbour, M., B. Pavlik, F. Drysdale, and S. Lindstrom. 1993. California's Changing Landscape: Diversity and Conservation of California Vegetation. Calif. Native Plant Society, Sacramento, California.

Barrett, S.A. and E.W. Gifford. 1933. Miwok Material Culture. Yosemite Natural History Association, Inc., Yosemite National Park, California

Bean, L.J. and K.S. Saubel. 1972. Temalpakh-Cahuilla Indian Knowledge and Usage of Plants. Malki Museum Press, Morongo Indian Reservation, Calif.

Brill, S. 1994. Identifying and Harvesting Edible and Medicinal Plants in Wild (and Not so Wild) Places. Hearst Books, New York.

Brown, T. 1985. Tom Brown's Guide to Wild Edible and Medicinal Plants. Berkeley Books, New York.

Callegari, J. And K. Durand. 1977. Wild Edible and Medicinal Plants of California. El Cerrito, CA

Chamberlain, L.S. 1901. Plants used by the Indians of Eastern North America. American Naturalist, 35:1-10.

Chesnut, V.K. 1902. Plants used by the Indians of Mendocino County, California. Contr U.S. Nat Herb 7:295-408.

Clarke, C.B. 1977. Edible and Useful Plants of California. Univ. Of Calif. Press, Berkeley, CA

Coffey, T. 1993. The History and Folklore of North American Wildflowers. Facts on File, Inc., New York

Coon, N. 1974. The dictionary of useful plants. Rodale Press/Book Division, Emmaus, PA

Coulter, J. and A. Nelson. 1909. New Manual of Botany of the Central Rocky Mountains (Vascular Plants) American Book Company

Coville, F.V. 1897. Notes on the plants used by the Klammath Indians of Oregon. Contrib. To the U.S. Nat. Herbarium, 5(2).

Craighead, J.J., F.C. Craighead, and R.J. Davis. 1963. A Field Guide to the Rocky Mountain Wildflowers. Boston: Houghton Mifflin Company.

Culpeper, N. 1972. English Physician and Complete Herbal. Arranged for use as a First Aid Herbal, by leyel, C.F. No. Hollywood, California. Wilshire Book Co.

Darlington, W. 1859. American Weeds and Useful Plants. New York, A.O. Moore

Dawson, R. 1985. Nature Bound. OMNIgraphics Ltd., Boise, Idaho

Densmore, F. 1974. How Indians Use Wild Plants for Food, Medicine, and Crafts. Dover Publications, New York.

Doebley, J.F. 1984. "Seeds" of Wild Grasses: A Major Food of Southwestern Indians. Economic Botany 38(I):52-64.

Duke, J.A. 1985. Handbook of Medicinal Plants. CRC Press, Boca Raton

Ebeling, W. 1986. Handbook of Indian Foods and Fibers of Arid America. Univ. of California Press, Berkeley, Los Angeles, London

Elias, T.S. and P.A. Dykeman. 1982. A Field Guide to North American Edible Wild Plants. Outdoor Life Books, New York.

Elliott, D.B. 1976. Roots: An Underground Botany and Foragers Guide. The Chatham Press, Old Greenwich, Conn.

Erichsen-Brown, C. 1979. Medicinal and Other Uses of North American Plants. Dover Publications, Inc., New York.

Farris, G. 1980. A re-assessment of the nutritional value of *Pinus monophylla*. Jour. of California and Great Basin Anthropology 2(1): 132-36.

Foster S. and J.A. Duke. 1990. Eastern/Central Medicinal Plants. Peterson Field Guides, Houghton Mifflin Company, New York

Frye, T.C. 1934. Ferns of the Northwest. Metropolitan Press, Portland, Oregon

Gibbons, E. 1962. Stalking the Wild Asparagus. New York: David McKay Co.

Gibbons, E. 1966. Stalking the Healthful Herbs. David McKay, New York.

Grillos, S.J. 1966. Ferns and Fern Allies of California. Univ. Of Calif. Press, Berkeley, CA

Gunther, E. 1973. Ethnobotany of Western Washington, revised edition. University of Washington Press, Seattle and London.

Hall, A. 1973. The Wild Food Trail Guide. Holt, Rinehart, and Winston.

Hardin, J.W. and J.M. Arena. 1974. Human Poisoning from Native and Cultivated Plants. Duke University Press, Durham, North Carolina.

Harrington, H.D. 1967. Edible Native Plants of the Rocky Mountains. Albuquerque: University of New Mexico Press.

Hart, J.A. 1976. Montana - Native Plants and Early Peoples, Montana Historical Society, Helena, MT.

Harvard, V. 1895. The Food Plants of North American Indians. Bulletin of Torrey Botanical Club 22:98-123.

Jacobson, C.A. 1915. "Water Hemlock (*Cicuta*)." Nevada Agricultural Experiment Station, Technical Bulleton #81.

Kavash, B. 1979. Native Harvests: recipes and botanicals of the American Indian. First Vintage Books, A

Division of Random House, NY.

Kindscher, K. 1987. Edible Wild Plants of the Prairie. University Press of Kansas.

Kingsbury, J.M.1964. Poisonous Plants of the United States and Canada. Englewood Cliffs, N.J.: Prentice-Hall.

Kingsbury, J.M. 1965. Deadly Harvest: A Guide to Common Poisonous Plants. New York: Holt, Rinehart, and Winston.

Kirk, D.R. 1970. Wild Edible Plants of the Western United States. Healdsburg, Calif.: Naturegraph Publishers.

Krochmal, A. and C. Krochmal. 1973. A Guide to the Medicinal Plants of the United States. Quadrangle/New York Times Book Company.

Krochmal, A, S. Paur, and P. Duisberg. 1951. Useful Native Plants in the American Deserts. Economic Botany 8(1):3-20.

Lewis, W.H. and M. Elvin Lewis. 1977. Medical Botany. John Wiley and Sons, New York.

Life-Support Technology, Inc. 1963. Foods in the Wilderness.

Lust, J.B. 1987. The Herb Book, 20th Edition. Bantam Books, New York

McHarg, I.L. 1969. Design with Nature. Garden City, New York. Doubleday

Medsger, O.P. 1974. Edible Wild Plants. Collier-Macmillan Publishers, New York

Merrill, R.E. 1923. Plants Used in Basketry by the California Indians. UC-PAAE 20:215-242.

Millspaugh, C.F. 1974. American Medicinal Plants, An Illustrated and Descriptive Guide to Plants Indigenous to and Naturalized in the United States Which are Used in Medicine. Dover Publications, New York.

Moerman, D.E. 1977. American Medical Ethnobotany: a reference dictionary. Garland Publishing, Inc. New York and London.

Moore, M. 1979. Medicinal Plants of the Mountain West. Museum of New Mexico Press, Santa Fe, NM

Moore, M. 1989. Medicinal Plants of the Desert and Canyon West. Museum of New Mexico Press, Santa Fe, NM

Morton, J. 1963. Principal Wild Food Plants of the United States, excluding Alaska and Hawaii. Economic Botany 17:319-330.

Muenscher, W.C. 1939. Poisonous Plants of the United States. The MacMillan Company, New York

Murphey, E.V.A. 1990. Indian Uses of Native Plants. Meyerbooks, Glenwood, IL.

Newberry, J.S. 1888. Food and Fiber Plants of the North American Indians. Popular Scientific Monthly 32:31-46.

Niehaus, T.F. 1974. Sierra Wildflowers: Mt. Lassen to Kern Canyon. University of California Press, Berkeley, California.

Olsen, L.D. 1990. Outdoor Survival Skills. Brigham University Press, Provo, Utah.

Palmer, E. 1878. Plants Used by the Indians of the United States. The American Naturalist 12:593-606 (Sept.) and 646-655 (Oct.).

Pfeiffer, N.E. 1922. Monograph of the Isoetaceae. Ann. Mo. Bot. Gard. 9:79-232.

Pojar, J. and A. MacKinnon, (eds.) 1994. Plants of the Pacific Northwest Coast. Lone Pine Publishing, Washington.

Schofield, J.J. 1989. Discovering Wild Plants: Alaska, Western Canada, the Northwest. Alaska Northwest Books, Seattle, WA

Scully, V. 1970. A Treasury of American Indian Herbs: Their Lore and Their Use for Food, Drugs, and Medicine. Crown Publishers, Inc., New York

Smith, C.E. 1973. Man and His Foods: studies in the ethnobotany and nutrition - contemporary, primitive, and prehistoric non-European diets. The Univ. Of Alabama Press, Alabama.

Snow, C.R. 1935. Vegetables of the Alaska Wilderness. The Alaska Sportsman 1(4):6-8

Strike, S.S. 1994. Etnobotany of the California Indians, vol. 2. Aboriginal Uses of California's Indigenous Plants. Koeltz Scientific Books, Champaign, Illinois

Sweet, M. 1962. Common Edible and Useful Plants of the West. Naturegraph Publishers, Inc., Healdsburg, Calif

Teit, J.A. 1930. Ethnobotany of the Thompson Indians of British Columbia. U.S. Government Printing Office, Washington, D.C.

Thompson, S. and M. Thompson. 1972. Wild Plant Foods of the Sierra. Dagwood Press,

Tilford, G.L. 1997. Edible and Medicinal Plants of the West. Mountain Press Publishing Company, Missoula, Montana.

Train, P., J.R. Henriches, and W.A. Archer. 1957. Medicinal Uses of Plants by Indian Tribes of Nevada. Quarterman Publication, Lawrence, Mass.

Tull, D. 1987. A Practical Guide to Edible and Useful Plants. Texas Monthly Press, Austin, TX

Turner, N. And H.V. Kuhnlein. 1991. Traditional Plant Foods of Canadian Indigenous Peoples. Gordon and Breach Science Publishers, Philadelphia, PA.

Turner, N.J. and A.F. Szczawinski. 1991. Common Poisonous Plants and Mushrooms of North America. Timber Press, Portland, OR

Van Etten, C.H., R.W. Miller, I.A. Wolff, and Q. Jones. 1963. Amino Acid composition of seeds from 200 angiosperm plant species. J. Agr. And Food Chem. 11(5):399-410.

Vestal, P.A. 1952. Ethnobotany of the Ramah Navaho. Papers of the Peabody Museum of American Archeology and Ethology, Harvard Univ. Vol. XL-No. 4

Vizgirdas, R.S. 2002. Useful Plants of Idaho. Idaho State University Press.

Webster, J. 1980. Fungi, 2nd Edition. Cambridge University Press.

Weedon, N.F. 1996. A Sierra Nevada Flora, 4th ed. Wilderness Press, Berkeley, California.

Willard, T. 1992. Edible and Medicinal Plants of the Rocky Mountains and Neighboring Territories. Wild

Rose College of Natural Healing, Ltd., Alberta, Canada

Wyman, L.C. and S.K. Harris. 1941. Navajo Indian Medical Ethnobotany. University of New Mexico Press, Albuquerque, New Mexico.

GLOSSARY

TO MEDICAL TERMS

Acrid. Sharp, irritating or biting to the taste.

Alkaloid. A nitrogen containing, slightly alkaline substance that is often poisonous.

Alterative. A substance that gradually restores the normal functions of the body.

Analgesic. Relieves pain.

Anaphrodisiac. An agent that reduces sexual desire or potency

Anesthetic. A substance that produces anesthesia

Anodyne. Helps to quiet or relieve pain.

Antibiotic. Helps to destroy pathogenic action of microbes.

Anti-inflammatory. To reduce or neutralize inflammation.

Anti-microbial. Something that inhibits the growth or multiplication of microorganisms, or kills them

Anti-pyretic. Against fever

Antiscorbutic. To be used against scurvy.

Antiseptic. Prevents infection.

Antispasmodic. Relieves or cures spasms or irregular and painful action of the muscles (e.g., epilepsy).

Anti-syphilitic. Relieves or cures venereal disease

Anti-viral. An agent used against viruses

Astringent. To shrink or bind tissue.

Bitters. Sharp acrid or biting medicines, prescribed to stimulate an appetite.

Cardiacs. To have an effect on the heart.

Carminatives. To dispel flatulency or griping pains of the stomach and bowels.

Cathartics. To stimulate the action of the bowels, a purgative.

Decoction. The essence of a plant extracted by boiling it down.

Demulcent. To have a soothing or emollient effect on inflamed surfaces.

Dermititis. Inflammation of the skin

Diaphoretics. To promote or increase perspiration.

Diuretics. To increase the flow of urine, by acting on the kidneys.

Emetics. Something that causes vomiting.

Emollients. To have a soothing and softening effect on the body tissues.

Expectorants. To cause an increase in expectoration, promoting the excretion of mucous from the chest.

Febrifuges. Help reduce or control fevers.

Glycocides. Any of the numerous acetal derivatives of sugars that on hydrolysis yield a sugar

Hydrocarbons. An organic compound containing one hydrogen and one carbon and often occurring in petroleum, natural gas, and coal.

Infusion. A preparation made by soaking a plant in hot water ("tea").

Insecticidal. Used against insects

Laxatives. Used to loosen the bowels and relieves constipation.

Liniment. A liquid or semi-liquid preparation of a herb to relieve skin irritation and muscle pain.

Mucilaginous. Something that resembles or contains mucilage (slimy).

Narcotics. Diminish the action of the nervous and vascular systems, causing drowsiness, lethargy, stupor, and insensibility.

Nervines. Soothe and calm the nerves, restoring them to a natural state

Nitrates. A salt or ester of nitric acid

Nutritives. Nourish the body, promoting growth or health.

Panacea. A "cure-all"

Parch. To toast or scorch with heat.

Photosensitivity. Being sensitive to light

Poultice. A moist, usually warm or hot mass of plant material applied to the skin, or with a cloth between the skin and plant material, to effect a medicinal action.

Purgatives. To evacuate the bowels, but more forcefully than a laxative.

Salve. A healing ointment

Saponin. A glycoside in plants that when shaken with water has a foaming or soapy action.

Sedative. Something used to lessen nervous excitement, irritation, and pain.

Selenium. A non-metalic element that resembles sulfur chemically.

Stimulants. Something produces energy.

Styptics. Helps control bleeding by contracting the tissues or blood vessels.

Sudorfic. To produceprofuse and visible sweating when taken hot

Tincture. A diluted alcohol solution of plant parts.

Tonic. To invigorate or stimulate, producing a feeling of well-being or strength.

Topical. Local application.

Vasoconstrictor. An agent that causes the blood vessels to constrict.

Vasodialator. An agent that causes the blood vessels to dialate.

Vermifuges. To expel worms or other parasites from the body.

Volatile. Readily vaporizes at low temperatures

Vulneraries. Used in the healing of wounds.

TO BOTANICAL TERMS

Achene. Small, dry and hard, 1 celled, 1-seeded fruit that is indehiscent

Acute. Sharp pointed

Alpine. Occurring above treeline in the mountains.

Alternate. Arranged with one structure (e.g., leaf, flowers, stem, etc.) per node.

Angiosperm. A plant producing flowers and bearing ovules (seeds) in an ovary (fruit).

Annual. A plant, usually with a slender taproot, completing its life cycle in a single growing season.

Anther. The pollen bearing portion of the stamen.

Aquatic. Growing in water.

Areola. Small defined area on a surface of a cactus that bears the spines

Aromatic. Having a strong, usually agreeable odor.

Asexual. Without sex

Awn. A slender bristle-like organ

Axil. The upper angle formed by a leaf or branch with the stem

Banner. The upper petal of a papilionaceous flower.

Basal. At the base

Beak. A prolonged, slender, and tapering projection.

Berry. A fleshy or pulpy fruit developed from a single ovary with more than one seed, such as a grape or blueberry.

Biennial. A plant completing its life cycle in two growing seasons; usually forming a basal rosette the first season and flowering the second.

Borne. Produced or arising from

Bract. A leaf subtending a flower or flower cluster.

Bulb. An underground organ constituted mostly of fleshy storage leaves and scale covered (e.g., onion).

Calyx. A flower's sepals considered as a unit.

Campanulate. Bell-shaped

Capillary. Hair- or thread-like.

Capsule. A dry dehiscent fruit, composed of more than one carpel

Carpel. A modified leaf forming the ovary.

Catkin. In plants such as willows, birches and alders, the elongated, pendulous or conelike flower cluster with minute flowers that lack, or almost, the petals and sepals.

Caudex. Thickened base of some perennial herbs

Clasping. Partly surrounding the stem.

Clavate. Club-shaped

Claw. The narrow or stalk-like base of some petals

Coma. The tufts of hairs at the ends of some seeds

Compound Leaf. One which is divided into two or more distinct leaflets.

Cone. A dense cluster of modified, leaflike organs bearing pollen, spores, or seeds as in horsetails, clubmosses, and conifers (e.g., pinecone).

Conifer. A cone bearing tree.

Coniferous. Having cones or strobili.

Corm. A bulb-like underground thickening of the stem.

Corolla. The petals considered as a unit, usually brightly colored.

Corymb. Convex or flat-topped flower cluster of the racemose type, with the pedicels arising from a different point on the axis.

Cruciform. Cross-shaped (e.g., the position of petals in the mustard family: Brassicaceae).

Cyathium. The inflorescence in the genus *Euphorbia* (Family Euphorbiaceae), that consists of unisexual flowers crowded within a cup-like involucre.

Cyme. A flower cluster, often flat-topped or convex, in which the central or terminal flower blooms the earliest.

Deciduous. Falling off once a year, usually at the end of a growing season.

Dehiscent. Opening to emit the contents

Dichotomous. Forking regularly by pairs

Discoid. A flowering head without ray flowers

Disk. In the sunflower family (Asteraceae), the central portion of the head that gives rise to the disk flowers.

Disk flowers. In the sunflower family (Asteraceae), the flowers with slender, tubular corollas at the central part of the head.

Divided. Cut or lobed to the base or to a midrib.

Drupe. A fleshy 1-seeded fruit (e.g., cherry).

Emergent. With the lower portion in water, and the upper portion extending out.

Endemic. Found only within a limited geographic area.

Entire. With margins not cut, cleft, or otherwise toothed.

Ephemeral. Lasting for only a short time.

Evergreen. Retaining leaves through the winter.

Exotic. Not native, but introduced from somewhere else.

Fascicle. Small cluster of leaves, flowers, etc.

Family. Group of related genera.

Flower. The reproductive portion of the plant consisting of stamens, pistils, or both, and including the petals, sepals, or both.

Foliage. The leaves of the plant, collectively

Follicle. A fruit consisting of a single carpel, dehiscing by the ventral suture

Forb. A non-grasslike herbaceous plant

Frond. The leaf of a fern.

Fruit. The mature ovary, that includes the attached external structures and enclosed seeds.

Genus. A grouping of related species. The plural, of which is genera.

Glabrous. Devoid of hairs

Gland. A secreting cell or group of cells

Glandular. Having glands, usually hairs.

Glaucous. Covered or whitened with a bloom

Glochid. A barbed hair or bristle (e.g., as in the fine hairs of Opuntia).

Gymnosperm. A member of the plant group that characterized as having ovules not enclosed in an ovary (e.g., pines, spruces, firs, junipers).

Habit. The general appearance or growth form of a plant.

Habitat. The environmental conditions or kind of place in which a plant grows.

Head. A type of inflorescence with mostly sessile flowers densely set on a very short axis or disk, thereby having a round outline.   The terminal collection of flowers surrounded by an involucre, as in the sunflower family (Asteraceae).

Herb. A plant with the aerial portion being non-woody, dying back to the ground at the end of the growing season.

Herbaceous. Not woody, dying back at the end of the growing season.

Host. The plants from which the parasite obtains nutrients (e.g., mistletoe)

Immersed. Growing under water.

Indehiscent. Not splitting open

Inferior. Lower or below

Inferior ovary. Ovary positioned below the base of other flower parts

Inflated. Turgid and bladdery

Inflorescence. The flowering part of plants; the arrangement of flowers

Inner bark. The cambium layer

Irregular. A flower where one or more of the organs are unlike the rest

Involucre. A whorl of bracts subtending a flower or flower cluster.

Latex The milky sap of certain plants.

Leaflet. One of the divisions of a compound leaf.

Legume. A simple dry fruit that is dehiscent along both sutures

Ligulate flower. The same as a ray flower in the Sunflower Family

Linear. Long and narrow

Many. For botanical purposes, numbering more than ten.

Meal. Pertaining to flour

-merous. Parted, having sections, as a 5-merous flower has 5 petals and 5 sepals.

Montane. Of or pertaining to the mountains.

Naturalized. Plants introduced from somewhere else, and now established.

Node. A joint or point of origin for leaves or branches.

Numerous. In botanical terms, more than ten.

Nut. A dry, hard walled and indehiscent fruit, usually with one seed.

Nutlet. A small nut.

Opposite. Nodes having two leaves or branches directly across from each other.

Orbicular. Circular in outline

Pappus. In the sunflower family, the highly modified calyx composed of scales, bristles, awns, or short crown at the tip of the achene.

Parasitic. Growing on and deriving nourishment from another living plant.

Parted. Cleft almost to the base

Pedicel. Stalk of a single flower

Perennial. Plant with the potential to live more than 2 years.

Perianth. Corolla and calyx considered collectively.

Petal. A member of the whorl of floral organs, just interior the sepals and below the stamens.

Petaloid. Brightly colored and petal-like.

Petiole. Leaf stalk

Pinnate. Having a main central axis with secondary branches or units arranged in two lines on either side of the central axis.

Pinole. The flour from various seeds and grains mixed together.

Pistil. Organ formed from the combination of the stigma, style, and ovary.

Plumose. Feathery and soft.

Pod. Any kind of dry, dehiscent fruit, particularly in the pea family (Fabaceae).

Pollen. Dustlike cells produced in the anther.

Pollinium. A mass of waxy or coherent pollen grains (e.g., notably in the Asclepiadaceae and Orchidaceae families).

Potherb. Herb that is boiled and eaten as a vegetable

Prostrate. Lying flat on the ground

Pubescent. Covered with hairs

Raceme. Inflorescence with one main axis and subequal primary branches each bearing one flower.

Ray flower. In the sunflower family (Asteraceae), the straplike flowers attached to the disk.

Receptacle. End of a flower stalk that bears the floral organs

Regular. Having members of each part alike in size and shape

Rhizomatous. Possessing rhizomes.

Rhizome. A creeping underground, usually horizontally oriented stem.

Root. The underground part of a plant.

Rootstalk. Underground, creeping stem.

Rosette. A dense, usually basal, cluster of leaves radiating in all directions from the stem.

Sagittate. Arrowhead shaped.

Salverform. A corolla with a long slender tube, abruptly flaring into a circular limb.

Samara. A winged fruit that does not split at maturity.

Saprophyte. A plant with little or no chlorophyll that obtains nutrients from dead organic matter by a root association with a fungus.

Schizocarp. Dry indehiscent fruit that splits into separate one-seeded segments at maturity.

Seed. A mature ovule which following germination gives rise to a new plant.

Seed cone. The female seed producing cone of conifers.

Sepal. One of a whorl of typically green or greenish, leaflike, floral organs originating below the petals.

Septum. A partition that seperates the locules of an ovary.

Serrate. Sharply toothed edges

Sessile. Lacking a stalk.

Shoot. Young stem or branch

Shrub. A woody plant, sometimes only at the base, and generally with several stems originating from the base.

Silicle. Dry, dehiscent fruit of the Mustard Family, typically less tha twice as long as wide

Silique. Dry, dehiscent fruit of the Mustard Family, typically more than twice as long as wide.

Spatulate. Spatula shaped; having a long narrow base and a widened, roundish tip.

Spore. A single cell or a small group of undifferentiated cells, each capable of producing a plant

Spur. A hollow slender sac-like extension of some part of the flower (e.g., sepal in Delphinium or the petal of Viola).

Stamen. The male organ of a flower that produces pollen. It is composed of an anther and filament.

Stellate. Star shaped

Stem. The main axis of a plant.

Stipules. Appendages on each side of the base of certain leaves

Subalpine. Growing in the mountains below the alpine zone and above the montane zone.

Succulent. Thick, fleshy, and juicy.

Superior. Above

Superior ovary. Ovary positioned above the base of the other flower parts

Talus. Slope of rock rubble, usually at a cliff base.

Taproot. The primary plant root that considerably larger than any other root system branches.

Tepal. The perianth part when the perianth is not clearly differentiated into calyx or corolla.

Terrestrial. Growing on ground, not aquatic.

Three-ranked. Originating in threes from a common point or level.

Tree. Large woody plant with a single main stem or trunk.

Tuber. A swollen underground stem tip (e.g., potato).

Tubular. In the form of a tube or cylinder.

Umbel. Flower arrangement resembling an "umbrella"

Urn shaped. Ovoid and with a small opening at the tip.

Vascular plant. A plant having vascular tissue.

Vegetative. The portion of the plant not producing reproductive structures like cones or flowers.

Weed. n aggressive plant that colonizes disturbed habitats and cultivated lands. A plant out of place.

Whorled. Three or more similar structures (e.g., leaves, petals, bracts) encircling a node

Woolly. Having soft, curled or entangled hairs.

INDEX

www.ingramcontent.com/pod-product-compliance
Lightning Source LLC
Chambersburg PA
CBHW071409180526
45170CB00001B/20